Nelson Mathematics 7

Student Success Workbook

Workbook Authors
Sandy DiLena • Rod Yeager

Series Authors and Senior Consultants
Marian Small • Mary Lou Kestell

Senior Author
David Zimmer

Assessment Consultant
Damian Cooper

Authors
Bernard A. Beales • Maria Bodiam • Doug Duff
Robin Foster • Cathy Hall • Jack Hope • Chris Kirkpatrick
Beata Kroll Myhill • Geoff Suderman-Gladwell • Joyce Tonner

NELSON / EDUCATION

NELSON / EDUCATION

Nelson Mathematics 7
Student Success Workbook

Series Authors and Senior Consultants
Marian Small, Mary Lou Kestell

Senior Author
David Zimmer

Workbook Authors
Sandy DiLena, Rod Yeager

Senior Program Consultant
Joanne Simmons

Student Text Authors
Bernard A. Beales, Maria Bodiam, Doug Duff, Robin Foster, Cathy Hall, Jack Hope, Chris Kirkpatrick, Beata Kroll Myhill, Geoff Suderman-Gladwell, Joyce Tonner

Director of Publishing
Beverley Buxton

Publisher, Mathematics
Colin Garnham

Project Manager, K-8
David Spiegel

Developmental Editor
Megan Robinson

Executive Managing Editor, Development & Testing
Cheryl Turner

Executive Managing Editor, Production
Nicola Balfour

Senior Production Editor
Susan Aihoshi

Proofreader
Dianne Broad

Editorial Assistant
Amanda Davis

Senior Production Coordinator
Sharon Latta Paterson

Creative Director
Angela Cluer

Art Director
Ken Phipps

Art Management
Suzanne Peden

Interior and Cover Design
Suzanne Peden

Illustrator
Deborah Crowle

Cover Image
Martin Barraud/Stone/Getty Images

Composition
Janet Zanette, Nelson Gonzalez, Pam Clayton

Printer
Globus

Library and Archives Canada Cataloguing in Publication

Nelson mathematics 7. Student success workbook / Sandy Dilena, Rod Yeager.

ISBN 0-17-628312-9

1. Mathematics—Problems, exercises, etc. I. Yeager, Rod, 1949- II. Title. III. Title: Nelson mathematics seven.

QA107.2.N44 2004 Suppl. 3
510   C2005-901695-7

Contents

Chapter 5: 2-D Measurement

Chapter 6: Addition and Subtraction of Integers

Chapter 7: 2-D Geometry

Chapter 8: Variables, Expressions, and Equations

Using Multiples

▶ **GOAL** Determine common multiples and least common multiples.

Problem 1

A group of students wants to sell about 100 hot dogs.

- The buns come in packages of 8.

- The wieners come in packages of 12.

The students want to make complete hot dogs with no buns or wieners left over. How many packages should they buy?

MATH T...

multiple
the produc...
whole numb... ...h
as 8, when m... ...d
by any other wh...
number, such as 1,
3, 4, and so on

Use these steps to find the common multiple of 8 and 12 that is closest to 100.

Step 1: Complete the table.

Number of packages	1	2	3	4	5	6	7	8	9	10	11	12
Number of ...ns	8	16	24									
N...ber of wi...rs	12	24										

Step ...: (Circle) each number in the table that appears in both rows of numbers.
These circled numbers are the common multiples of 8 and 12.

Which circled number is closest to 100? _____

Step 3: How many packages of buns and wieners should the students buy to make about 100 hot dogs?

_____ packages of buns

_____ packages of wieners

Problem 2

What is the fewest number of hot dogs the students can make without wasting any buns or wieners?

Use these steps to find the least common multiple of 8 and 12.

Step 1: Use the circled numbers in the table on page 2 to write the common multiples of 8 and 12.

_____, _____, _____, _____

Step 2: What is the least common multiple? _____

Step 3: What is the fewest number of hot dogs the students can make without wasting any buns or wieners?

_____hot dogs

Here is a summary of how to find the least common multiple of two numbers, such as 5 and 8.

List the multiples of each number from least to greatest.

5, 10, 15, 20, 25, 30, 35, (40), 45, …

8, 16, 24, 32, (40), 48, …

Circle the smallest number that appears in both lists. The least common multiple of 5 and 8 is 40.

Reflecting

▶ How does making a list of the multiples of two numbers help to find the least common multiple?

_____

_____

_____

Practising

Text page 6 **3. a)** Complete the multiplication table.

X	1	2	3	4	5	6	7	8	9	10	11	12	13	14	15
2	2	4	6												
5	5	10													
6	6														

b) (Circle) the common multiple(s) of 2, 5, and 6 in the table.

c) What is the **LCM** of 2, 5, and 6? _____

> **MATH TERM**
>
> **LCM**
> a short way of writing **L**east **C**ommon **M**ultiple

8. You expect to sell between 80 and 100 hamburgers at a baseball tournament.

• Hamburger buns are sold in packages of 6.

• Meat patties are sold in packages of 8.

You want to make complete hamburgers with no buns or meat patties left over. How many packages should you buy?

a) Complete the table to help solve the problem.

Number of packages	1	2	3	4	5	6	7	8	9	10	11	12	13	14	15	16	17
Number of buns	6																
Number of patties	8																

b) I should buy the following:

_____ packages of buns

_____ packages of meat patties

Connect Your Work

Here is a quick method you can use to determine if a number is a common multiple of two numbers:

If one number divides equally into another number, the greater number is a multiple of the lesser number.

For example, if 15 is a multiple of 3, then 3 divides into 15 with no remainder.

$15 \div 3 = 5$ remainder 0

Since there is no remainder, 15 is a multiple of 3.

If 15 is a multiple of 5, then 5 divides into 15 with no remainder. Check to see if this is true.

$15 \div 5 =$ _____ remainder _____

Is 15 a common multiple of 3 and 5? _____

Hint

You can use counters or draw pictures to model $15 \div 5$.

Practise this idea.

Text page 7

9. Is 135 a common multiple of 3 and 5?
   Show your work.

END

A Factoring Experiment

You will need
- coloured pencils

- a calculator

▶ **GOAL** Identify factors of numbers.

Problem

This grid represents a tiled floor. ▶

There are 144 tiles.
The area of each tile is 1 m².
There are 12 tiles on each side.
12 tiles × 12 tiles = 144 tiles

12 m

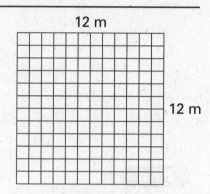

12 m

A. This grid shows what the floor would look like using larger tiles. ▶

Each tile measures 2 m by 2 m. The area of each tile is 4 m².

How many 2 m by 2 m tiles are there on each side?

_____

How many tiles are there altogether? _____

Write a multiplication sentence to describe the total number of tiles.

_____ tiles × _____ tiles = _____ total tiles

12 m

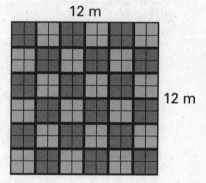

12 m

Hint

Use two different colours so you can see the separate tiles.

B. On this grid, colour in square tiles that measure 3 m by 3 m. ▶
The first tile is shaded for you.

12 m

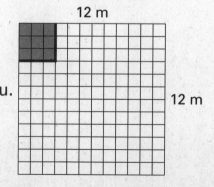

12 m

C. Complete this table. Look for a pattern.

Hint

Use the last grid from page 6 to help you complete the third row of the table.

Side length of one tile (m)	Area of one tile (m²)	Number of tiles per side	Total number of tiles
1	1	12	144
2	4	6	36
3			
4			
6			
12			

D. Why do you always get 12 when you multiply the side length of one tile by the number of tiles per side?

_____

_____

_____

E. Look at each row in the table. Explain why you can multiply a number in the third column by itself to get the number in the fourth column.

_____

_____

_____

Hint

Think about what you found out in Part A.

Reflecting

▶ How does knowing the factors of 12 help you find tiles that fit?

_____

_____

_____

_____

1.3 Factoring

Text page 10

You will need
- linking cubes (optional)

▶ **GOAL** Determine factors, common factors, and greatest common factors of whole numbers.

Problem

Find the common factors and greatest common factor of 24 and 18.

You can make rectangles with an area of 24 units² to find all the factors of 24.

MATH TERM

factor
one of the numbers you multiply in a multiplication operation
24 × 1 = 24
↑   ↑
factor factor

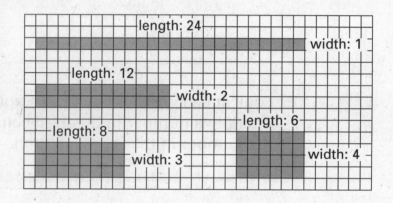

The lengths and widths show all the factors of 24.

Hint

Remember, length × width = area.

Length	24	12	8	6
Width	1	2	3	4

Write the factors of 24 in order from least to greatest.

___1___, _____, _____, _____, _____, _____, _____, ___24___

Hint

Use linking cubes to help you make all the rectangles with an area of 18.

Use these steps to find the factors of 18.

Step 1: Draw all the rectangles with an area of 18 units^2 on the grid. The first one is done for you.

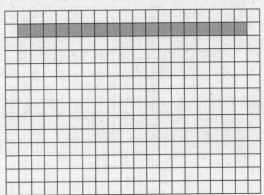

Step 2: Write the lengths and widths in this table.

Length	18		
Width	1		

Step 3: Write the factors of 18 in order from least to greatest.

___1___, _____, _____, _____, _____, ___18___

Use these steps to find the greatest common factor of 24 and 18.

Step 4: Write the common factors of 24 and 18. Remember, the factors of 24 are
1, 2, 3, 4, 6, 8, 12, 24

_____, _____, _____, _____

Step 5: Which number is the **greatest common factor**?

_____

Reflecting

▶ How does ordering factors from least to greatest help you find the greatest common factor?

_____

_____

Practising

7. a) Draw all the rectangles with areas of 12 and 18.

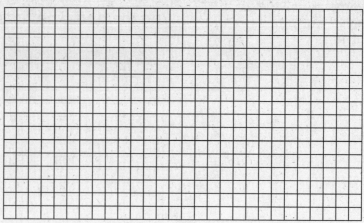

b) Write the length and width of each rectangle in the tables.

Rectangles for 12	Length	12		
	Width	1		

Rectangles for 18	Length	18		
	Width	1		

c) List the factors for both numbers in order from least to greatest.

factors of 12: _____

factors of 18: _____

(Circle) the factors that appear in both lists.

d) List the common factors of 12 and 18.

_____, _____, _____, _____

MATH TERM

GCF
a short way of
writing **G**reatest
Common **F**actor

e) What is the **GCF** of 12 and 18? _____

8. a) Write the lengths and widths of all rectangles with an area of 8.

Rectangles for 8	Length		
	Width		

Write the lengths and widths of all rectangles with an area of 10.

Rectangles for 10	Length		
	Width		

List the factors of 8 and 10 from least to greatest.

factors of 8: 1, _____

factors of 10: 1, _____

Circle the common factors.

What is the GCF of 8 and 10? _____

b) List the factors of 3 and 12 from least to greatest.

factors of 3: 1, _____

factors of 12: 1, _____

Circle the common factors.

What is the GCF of 3 and 12? _____

c) List the factors of 6 and 24 from least to greatest.

factors of 6: 1, _____

factors of 24: 1, _____

Circle the common factors.

What is the GCF of 6 and 24? _____

Exploring Divisibility

You will need
- a calculator
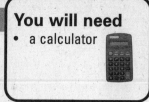

▶ **GOAL** Explore divisibility rules.

This model shows 10 divided into groups of 3.

 ◀— remainder of 1

10 is **not** divisible by 3 because there is a remainder of 1.

(Circle) all groups of 3 on the model below to show 18 divided by 3.

   Write the remainder. _____

18 is divisible by 3 because there is a remainder of 0.

Problem

Use these steps to find out if 138 is divisible by 3.

Here is a model of 138 as 1 hundred + 3 tens + 8 ones.

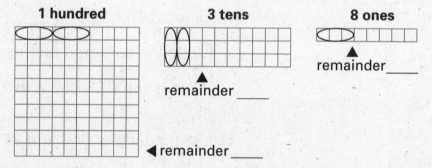

1 hundred   **3 tens**   **8 ones**

remainder ____ (8 ones)

remainder ____ (3 tens)

◀ remainder ____ (1 hundred)

Step 1: (Circle) all groups of 3 on each part of the model.

Step 2: Can you make a group of 3 out of all the remainders? _____

What is the total remainder now? _____

Step 3: Is 138 divisible by 3? How do you know?

_____

Here's another way to find out if 138 is divisible by 3.

Step 4: Add the digits of 138.

$$1 + 3 + 8 = \underline{\hspace{2cm}}$$

Is the sum of the digits of 138 divisible by 3?

_____

Step 5: Are the sum of the digits of 138 and the number 138 both divisible by 3?

_____

See if this helps you find out if other numbers are divisible by 3. Try 126.

Step 6: Divide 126 by 3.

$$126 \div 3 = \underline{\hspace{2cm}} \text{ remainder } \underline{\hspace{2cm}}$$

Is 126 divisible by 3? _____

Step 7: Add the digits of 126.

$$1 + 2 + 6 = \underline{\hspace{2cm}}$$

Is the sum of the digits of 126 divisible by 3?

_____

Step 8: Are the sum of the digits of 126 and the number 126 both divisible by 3?

_____

Reflecting

▶ How can you predict if a number is divisible by 3?

_____

_____

_____

1.5 Powers

Text page 16

▶ **GOAL** Use powers to represent repeated multiplication.

Use these steps to write a multiplication sentence as a power.

Step 1: Fold a piece of paper in half. Unfold the paper and count the number of sections.

1 fold	2

Step 2: Refold the paper along the same fold. Then fold it in half again. Count the number of sections.

1	2
3	4

Step 3: Continue folding the paper in half. Complete the table.

Number of folds	Number of sections	Repeated multiplication	Power
1	2	2	2^1
2	4	2×2	2^2
3	8	$2 \times 2 \times 2$	2^3
4		$2 \times \underline{\hspace{1cm}} \times \underline{\hspace{1cm}} \times \underline{\hspace{1cm}}$	2^4
5			
6			

Step 4: Look for a pattern in the |third column| of the table. Describe the pattern.

_____

_____

_____

Step 5: Continue the pattern.
Write the multiplication sentence for the 7th fold.

_____ = _____

A short way to write this multiplication sentence is 2^7. This is called a **power**.

Step 6: Look for a pattern in the powers in the fourth column of the table. Describe the pattern.

_____

_____

_____

Reflecting

▶ Is 2^7 double 2^6? Why?

_____

_____

Connect Your Work

1. Write the multiplication sentence for 3^5.

_____ = _____

2. Use a power to represent each of the following repeated multiplications. Then find the product. The first one is done for you.

   a) $2 \times 2 \times 2 \times 2 = 2^4$
   $\qquad\qquad\qquad\quad = 16$

   b) $5 \times 5 \times 5 \times 5 =$ _____
   $\qquad\qquad\qquad\quad =$ _____

   c) $10 \times 10 \times 10 \times 10 \times 10 \times 10 =$ _____
   $\qquad\qquad\qquad\qquad\qquad\qquad =$ _____

Practising

Text page 18 **6.** A lottery winner has two options on March 1.

Option 1: Take $500 000 cash immediately.

Option 2: Take the prize on March 12.
The prize is calculated by tripling the amount
each day. The prize starts with $3 on March 1.

a) Complete the table to determine which option is
the better deal.

Date	Repeated multiplication	Power	Prize ($)
March 1	3	3^1	3
March 2	3×3	3^2	9
March 3	$3 \times 3 \times 3$	3^3	27
March 4			
March 5			
March 6			
March 7			
March 8			
March 9			
March 10			
March 11			
March 12			

b) Which option is the better deal? _____

10. Complete the tables to find out how much money a May 1 lottery winner would collect on May 6 for each option.
Multiply the amount won each day by the number you started with to get the winnings the next day.

a)

Date	Repeated multiplication	Power	Prize ($)
May 1	5	5^1	5
May 2	5×5	5^2	25
May 3	$5 \times 5 \times 5$		
May 4			
May 5			
May 6			

b)

Date	Repeated multiplication	Power	Prize ($)
May 1	10	10^1	10
May 2	10×10	10^2	100
May 3	$10 \times 10 \times 10$		1000
May 4			
May 5			
May 6			

12. Fill in the missing parts in these multiplication sentences. The first one is done for you.

a) $36 = 6 \times 6$
$= 6^2$

b) $49 = \underline{} \times \underline{}$
$= \underline{}^2$

c) $81 = \underline{} \times \underline{}$
$= \underline{}^2$

d) $100 = \underline{} \times \underline{}$
$= \underline{}^2$

Square Roots

You will need
• a calculator

▶ **GOAL** Determine the square roots of perfect·squares.

Problem 1

MATH TERM

perfect square
a square with equal side lengths that are whole numbers

The floor mat in gymnastics is a **perfect square** with an area of 144 m².

This grid is a model of the mat. ▶

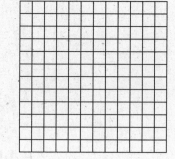

Use these steps to find out how the side length of the mat relates to its area.

Step 1: Count the squares to find each side length of the mat.

What is the side length of the mat? _____

Step 2: Multiply the side length by itself.

12 × 12 = _____

Is the product the same as the area of the mat?

_____

Rachel said, "I can find the side length of any square if I know the area. I only need to find a number that gives the area of the square when multiplied by itself."

Is Rachel correct? Why?

_____

_____

_____

Problem 2

MATH TERM

square root
the side length of a square is called the square's root or square root; $\sqrt{}$ is the symbol for square root

Hint

Remember, the side lengths of a perfect square are equal whole numbers.

Find the side length of a square that has an area of 49 cm². Then find the **square root** of 49.

Use guessing and testing to help find $\sqrt{49}$.

$5 \times 5 =$ _____   This means 5 is too _____.
                                                                              small/big

$10 \times 10 =$ _____   This means 10 is too _____.
                                                                              small/big

Is the square root of 49 closer to 5 or 10? _____

Try more guesses.

_____ $\times$ _____ $=$ _____

_____ $\times$ _____ $=$ _____

Keep guessing until you find $\sqrt{49} =$ _____.

Reflecting

▶ Why are there no perfect squares between 25 and 36?

_____

_____

_____

Practising

Text page 24

7.

11 cm

11 cm

a) Write the area of the square. _____

b) Complete. $\sqrt{} = 11$

TURN ➡

9. Sandra wondered if the square root of 169 is between 10 and 20.

• She used guessing and testing to find $\sqrt{169}$.

• First she tried 10. Then she tried 20.

Help Sandra find $\sqrt{169}$.

a) $10 \times 10 =$ _____   This means 10 is too _____.

b) $20 \times 20 =$ _____   This means 20 is too _____.

c) Is $\sqrt{169}$ closer to 10 or 20? _____

d) Try more guesses.

_____ $\times$ _____ = _____

_____ $\times$ _____ = _____

Keep guessing until you find $\sqrt{169} =$ _____.

11. The competition area in judo is a perfect square with an area of 256 m². Complete to find the side length of the area.

a) Suppose the side length of the mat is 10 m.

$10 \times 10 =$ _____   This means 10 is too _____.

b) Suppose the side length of the mat is 20 m.

$20 \times 20 =$ _____   This means 20 is too _____.

c) Is the side length of the judo mat closer to 10 m or 20 m?

_____

d) Keep guessing until you find $\sqrt{256} =$ _____.

Connect Your Work

This table contains all perfect squares with side lengths from 1 to 10 units. Complete the table.

Side length (unit)	Perfect square area (unit^2)	Square root $\sqrt{}$
1	$1 \times 1 = 1^2$ $= 1$	$\sqrt{1} = 1$
2	$2 \times 2 = 2^2$ $= 4$	$\sqrt{4} = 2$
3		
4		
5		
6		
7		
8		
9		
10		

END

Order of Operations

You will need
• a calculator

▶ **GOAL** Apply the rules for order of operations.

Problem 1

Anna's family ate hamburgers on their picnic.

• 5 people had 1 single-patty hamburger each.

• 4 people had 1 double-patty hamburger each.

• 2 people had 1 single-patty cheeseburger each.

How many patties did Anna's family eat?

Anna's Solution

Anna used an equation: $5 \times 1 + 4 \times 2 + 2 \times 1 = 20$. There were 11 people at the picnic and she knows that 11×2 is about 20, but most people ate only 1 patty. She knew her calculation was wrong.

Anna decided she had to use **BEDMAS** to solve her problem.

MATH TERM

BEDMAS
the rules for order of operations:
1. **B**rackets
2. **E**xponents
3. **D**ivide and **M**ultiply from left to right
4. **A**dd and **S**ubtract from left to right

Step 1: There are no brackets, so Anna went to Step 2.

Step 2: There are no exponents, so Anna went to Step 3.

Step 3: Anna multiplied.

$$\underline{5 \times 1} + \underline{4 \times 2} + \underline{2 \times 1}$$
$$= 5 + 8 + 2$$

Step 4: Anna added.

$$= \underline{5 + 8 + 2}$$
$$= 15$$

Anna's family ate 15 patties.

**Follow these steps to use BEDMAS to evaluate
$(5 + 2)^2 \times (2 + 2)$.**

Step 1: Check for brackets.
Underline the parts in brackets.
Add the parts in brackets from left to right.

$(5 + 2)^2 \times (2 + 2)$

= _____ × _____

Go to Step 2.

Step 2: Check for exponents.
Underline the parts with exponents.
Evaluate the power.

= _____ × _____

= _____ × _____

Go to Step 3.

Step 3: Check for division and multiplication.
Underline the multiplication.
Then multiply from left to right.
Use your calculator.

= 49 × 4

= _____

The problem is solved. There is no need to go to Step 4.

Reflecting

▶ Press the following keys on your calculator to see if
you get the same answer as Anna.

5 ⊠ 1 ⊞ 4 ⊠ 2 ⊞ 2 ⊠ 1 ⊟ _____ .

Does your calculator follow BEDMAS?
How do you know?

_____

_____

Practising

Text page 28

Hint

You may use your calculator for individual calculations.

Hint

Remember,
1. **B**rackets
2. **E**xponents
3. **D**ivide and
   Multiply from left to right
4. **A**dd and
   Subtract from left to right

6. Calculate. Underline the calculation for each step. Show all your work. The first one is done for you.

a) $\underline{(15 - 12)} \div 3 \times 2 - 1$

$= \underline{3 \div 3} \times 2 - 1$

$= \underline{1 \times 2} - 1$

$= \underline{2 - 1}$

$= 1$

c) $15 - (12 \div 3) \times 2 - 1$

$=$

$=$

$=$

b) $15 - 12 \div 3 \times (2 - 1)$

$=$

$=$

$=$

$=$

d) $(15 - 12 \div 3) \times 2 - 1$

$=$

$=$

$=$

$=$

e) Which answer in parts a), b), c), and d) gives the same answer as $15 - 12 \div 3 \times 2 - 1$?

_____

8. Calculate. Underline the calculation for each step. Show all your work. The first one is done for you.

a) $\underline{12 \times 9} + 3$

$= \underline{108 + 3}$

$= 111$

c) $5^2 + 5 - 4 \div 2$

$=$

$=$

$=$

$=$

b) $12 + 9 \times 3$

$=$

$=$

d) $(5 + 2^2 \times 3)^2 + 1$

$=$

$=$

$=$

$=$

$=$

12. Calculate. Underline the calculation for each step. Show all your work. The first one is done for you.

a) $3 \times 5 + \underline{10^3} \times 3$

$= \underline{3 \times 5} + \underline{1000 \times 3}$

$= 15 + 3000$

$= 3015$

c) $5^3 + 4 \div 2 - 5 \times 6$

$=$

$=$

$=$

$=$

b) $10 + (3^2 - 1) \div 2 - 4$

$=$

$=$

$=$

$=$

d) $4 \times (3 - 1 \times 3) + 8 \div 2^3$

$=$

$=$

$=$

$=$

$=$

Text page 29

16. Show how you could get two different answers when evaluating $125 \div 5 \div 5$ if there were no rules for order of operations.

$125 \div 5 \div 5$

$=$

$=$

$125 \div 5 \div 5$

$=$

$=$

END

Solve Problems by Using Power Patterns

▶ **GOAL** Use patterns to solve problems with powers.

Problem

Help Teresa find the last digit of 3^{15} without calculating the power.

1 Understand the Problem

Teresa understands, "I need to find the last digit of 3^{15} without calculating the value of the power. I know I can calculate the value of other powers. I don't need to know the value of 3^{15}, only the last digit."

2 Make a Plan

Teresa decides, "I am going to calculate the value of 3^1, 3^2, 3^3, 3^4, and so on. I will look for the patterns in the last digits. I will write the powers in a table so it is easy to see the pattern."

3 Carry Out the Plan

Step 1: Calculate these powers. Use a calculator.

$3^1 = $ _____ $3^2 = $ _____ $3^3 = $ _____ $3^4 = $ _____

$3^5 = $ _____ $3^6 = $ _____ $3^7 = $ _____

$3^8 = $ _____ $3^9 = $ _____ $3^{10} = $ _____

Step 2: List the last digit of each calculated power.

_____

Step 3: Describe the pattern of the last digits.

_____

Step 4: There are four numbers in the pattern so there should be four columns in the table.
Write the calculated powers from Step 1 in this table.

Column 1	Column 2	Column 3	Column 4
$3^1 =$ _____	$3^2 =$ _____	$3^3 =$ _____	$3^4 =$ _____
$3^5 =$ _____	$3^6 =$ _____	$3^7 =$ _____	$3^8 =$ _____
$3^9 =$ _____	$3^{10} =$ _____		

Step 5: Look at the exponents in each column of the table. What do you have to add to an exponent to get the next exponent in the same column? _____

Step 6: Use the pattern you found in Step 5 to determine what column 3^{15} belongs in. _____

Step 7: What is the last digit of the calculated powers in the column you named in Step 6? _____

Step 8: What is the last digit of the calculated power 3^{15}? _____

4 Look Back

Use a calculator to check what the last digit of 3^{15} is. Was your answer in Step 8 correct? _____

Reflecting

▶ How could you use the pattern to find the last digit of 3^{20}? What is the last digit of 3^{20}?

_____

_____

_____

TURN

Practising

Text page 32 **5.** Find the last digit of 2^{20} and 2^{30}.

a) Complete the table.

Column 1	Column 2	Column 3	Column 4
$2^1 = $ _____	$2^2 = $ _____	$2^3 = $ _____	$2^4 = $ _____
$2^5 = $ _____	$2^6 = $ _____	$2^7 = $ _____	$2^8 = $ _____
$2^9 = $ _____	$2^{10} = $ _____	$2^{11} = $ _____	$2^{12} = $ _____

b) What column would 2^{20} be in? _____
What is the last digit of the calculated power 2^{20}?
_____

c) What column would 2^{30} be in? _____
What is the last digit of the calculated power 2^{30}?
_____

7. Nathan calculated squares of numbers that end in 5.

a) Underline the last two digits in each calculated power.
What pattern do you see?

$15^2 = \ 2\underline{25}$
$25^2 = \ 625$
$35^2 = 1225$
$45^2 = 2025$
$55^2 = 3025$

_____

b) Nathan found a pattern for the first digits of the answers. Complete the table to show the pattern.

c) Use the patterns you found to calculate:

$65^2 = $ _____

$75^2 = $ _____

$85^2 = $ _____

First digits
$1 \times 2 = \ 2$
$2 \times 3 = \ 6$
$3 \times 4 = 12$
$4 \times 5 = 20$
$5 \times 6 = 30$
$6 \times$ __ $=$ ____
$7 \times$ __ $=$ ____
$8 \times$ __ $=$ ____

8. Find a power of 2 that is greater than 2^{100} and has a last digit of 4.

a) Complete the table. Use your answers from question 5 a).

Column 1	Column 2	Column 3	Column 4
$2^1 =$ _____	$2^2 =$ _____	$2^3 =$ _____	$2^4 =$ _____
$2^5 =$ _____	$2^6 =$ _____	$2^7 =$ _____	$2^8 =$ _____
$2^9 =$ _____	$2^{10} =$ _____	$2^{11} =$ _____	$2^{12} =$ _____

b) What column is the power in? _____

c) What is the power? _____

9. a) Use your calculator to multiply all the numbers from 1 to 10 together.

$1 \boxed{\times} 2 \boxed{\times} 3 \boxed{\times} 4 \boxed{\times} 5 \boxed{\times} 6 \boxed{\times} 7 \boxed{\times} 8 \boxed{\times} 9 \boxed{\times} 10$

= _____

Hint
Look at the factors of the product. Try multiplying any number by 10.

b) Why must the last digit of the product be 0?

_____

_____

_____

c) Without calculating, what is the last digit of the product of all the numbers from 1 to 99? _____ How do you know?

_____

_____

_____

Exploring Ratio Relationships

▶ **GOAL** Explore equivalent ratios.

The ratio of width to length for Rectangle 1 is **width : length = 2 : 3**.

The width, length, and ratio of Rectangle 1 are recorded in this table.

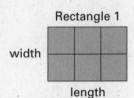

Rectangle 1

width

length

Rectangle	Width	Length	Width : Length
1	2	3	2 : 3
2			
3			
4			

Follow these instructions to draw similar rectangles for Rectangle 1.

A. Create Rectangle 2 by adding 2 units to the width and 3 units to the length. Shade and label the rectangle. Complete the next row of the table above.

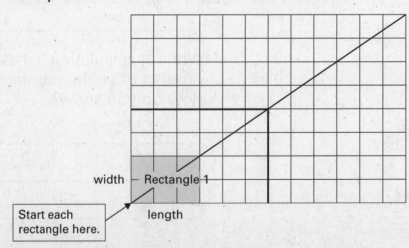

width — Rectangle 1

Start each rectangle here.

length

B. Shade Rectangle 3 and 4 in different colours. For each rectangle, add 2 units to the width and 3 units to the length of the previous rectangle. Label the rectangles "Rectangle 3" and "Rectangle 4."

C. Record the width, length, and ratio of each rectangle in the table on page 30.

D. The ratios (width : length) of similar rectangles are equivalent. Rectangles 1, 2, 3, and 4 are similar rectangles. So, the ratios in the table are equivalent. Each number in the first ratio can be multiplied by the same number to get the next ratio.
Show how this is true for Rectangles 1 and 2.

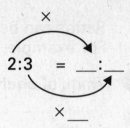

$$2:3 = \underline{\quad} : \underline{\quad}$$

MATH TERM

similar rectangle
a rectangle that has the same shape as another rectangle, but not necessarily the same size

Reflecting

▶ How much should you add to the width and length each time to draw a **similar rectangle** to the rectangle below? Use the grid to check your answer.

add to width:
_____

add to length:
_____

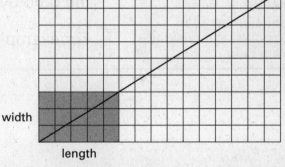

width

length

▶ Are the rectangles below similar? How do you know?

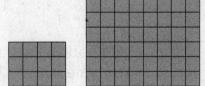

_____

_____

_____

Solving Ratio Problems

▶ **GOAL** Compare quantities using ratios, and write equivalent ratios.

Ratios can be written as fractions.
For example, 1:2 can be written as $\frac{1}{2}$.

Think of **equivalent ratios** as equivalent fractions.

MATH TERM

equivalent ratios
two or more ratios that show the same comparison; 1:2 and 2:4 are equivalent ratios

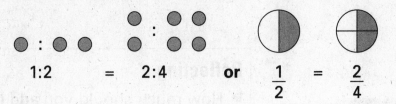

1:2   =   2:4   **or** $\frac{1}{2}$ = $\frac{2}{4}$

MATH TERM

term
the number that represents a quantity in a ratio

A. To create an equivalent ratio, multiply each **term** in the ratio by the same number.

For example:

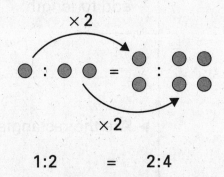

1:2   =   2:4

Use counters to model two more ratios that are equivalent to 1:2. Write the ratios below.

1:2 = 2:4 = _____:_____ = _____:_____

B. You can also divide each number in a ratio by the same number to create an equivalent ratio.

For example:

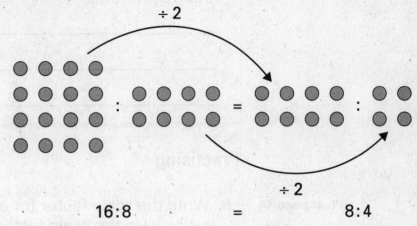

$$16:8 = 8:4$$

Use counters to model two more ratios that are equivalent to 16:8. Write the ratios below.

16:8 = 8:4 = _____:_____ = _____:_____

C. Write the **scale factor** for each pair of equivalent ratios. The first one is done for you.

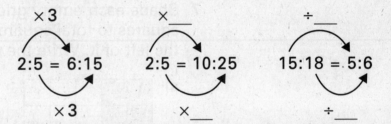

D. Write the scale factor for each pair of equivalent ratios. Then use the scale factor to find the missing term in the second ratio.

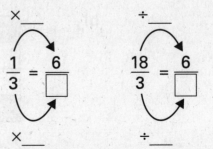

Reflecting

▶ How can you find the scale factor used to create an equivalent ratio?

_____

_____

_____

Practising

Text page 44

6. Write the scale factor for each pair of equivalent ratios. Use the scale factor to find the missing term.

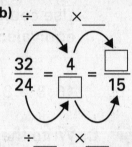

a)
$$\frac{3}{8} = \frac{6}{\square}$$

b)
$$\frac{32}{24} = \frac{4}{\square} = \frac{\square}{15}$$

7. Shade each empty grid so that the ratio of shaded squares to total squares is the same as on the grid to the left of it. Write the equivalent ratio below the grid.

a)
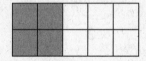

4 : 10   =   _____ : _____

b)

2 : 4   =   _____ : _____

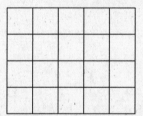

8. Write two equivalent ratios for each of the following ratios. Use a calculator.

a) 21 to 56 = _____ to _____ = _____ to _____

b) 6 : 54 = _____ : _____ = _____ : _____

c) $\dfrac{22}{55} = \dfrac{\square}{\square} = \dfrac{\square}{\square}$

Text page 45

9. The ratio of the number of shaded sections to the total number of sections is equivalent for all three diagrams.

a) Write the ratio below each diagram.

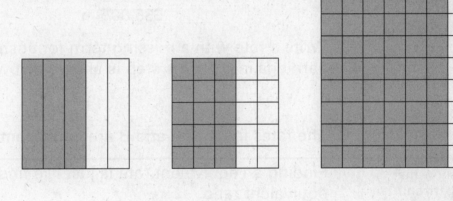

_____ : _____   _____ : _____   _____ : _____

b) Explain why each diagram represents the same ratio.

_____

_____

_____

Solving Rate Problems

▶ **GOAL** Determine equivalent rates to solve rate problems.

Problem 1

Josh worked 5 hours and earned $35.00.
How much does Josh make per hour?

Follow these instructions to solve the problem.

MATH TERM

rate
a way to compare
two quantities
measured in different
types of units

A. Write the **rate** for Josh's earnings in 5 h.
This step is done for you.

$$\$35.00/5 \text{ h}$$

B. Write a rate with a missing term for Josh's
earnings in 1 h. This step is shown below.

$$\$ \boxed{} /1 \text{ h}$$

MATH TERM

equivalent rates
rates that show the
same comparison;
for example,
$9.00/1 h and
$90.00/10 h

C. The rates in Parts A and B are **equivalent rates**.

Finding an equivalent rate is just like finding an
equivalent ratio.

Use the scale factor to find the missing term in the
equivalent rate.

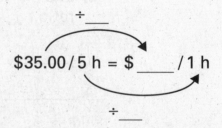

$$\$35.00 / 5 \text{ h} = \$ \underline{\quad} / 1 \text{ h}$$

Josh makes $\underline{\quad}/h.

If Sandra makes $8.00/h, how much money would she make if she worked 4 h?

Use a scale factor to find the missing term in the equivalent rate.

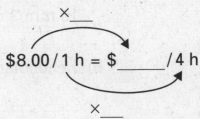

$8.00 / 1 h = $_____ / 4 h

Sandra would make $ _____ in 4 h.

Reflecting

▶ Why do you not make any more dollars per h when you make $35.00/5 h than when you make $7.00/h?

_____

▶ Why is it important to include the units when you say "$7.00/h?"

_____

_____

_____

▶ How are rates and ratios similar?

_____

_____

_____

▶ How are rates and ratios different?

_____

_____

_____

_____

Practising

Text page 48   **5.** Write each comparison as a rate.
The first one is done for you.

   a) There were 15 mm of rain over 3 days.

   15 mm/3 days

   b) 4 chocolate bars were on sale for $2.20.

   _____

   c) You save $14.00 in 1 week.

   _____

6. Write two equivalent rates for each comparison.
The first one is done for you.

   a) 5 goals in 10 games = 1 goal/2 games

   = 10 goals/20 games

   b) 10 km jogged in 60 min = _____ km/_____ min

   = _____ km/_____ min

   c) 6 pizzas eaten in 30 min = _____ pizzas/_____ min

   = _____ pizzas/_____ min

7. On a hike, Peter walked 28 km in 7 h.
What is his average hourly rate of walking?
Use a scale factor to find the missing term.

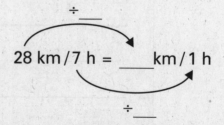

28 km/7 h = _____ km/1 h

Peter's average rate of walking is _____ km/h.

8. For each problem, write the **proportion**, then find the missing term. The first one is done for you.

a) 3 trucks have 54 wheels.
How many wheels do 6 trucks have?

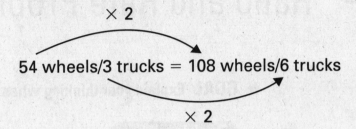

$\times 2$

54 wheels/3 trucks = 108 wheels/6 trucks

$\times 2$

b) In 5 h, your mom drove 400 km.
How many km did she drive in 1 h?

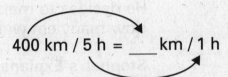

400 km / 5 h = ___ km / 1 h

c) 6 boxes contain 72 donuts.
How many donuts does 1 box contain?

13. Anita earns $72 every 6 h to fix bicycles.

a) How much does she earn in 1 h?

$72/6 h = ___ /1 h

b) How much money will she earn in 4 h?

___ /1 h = ___ /4 h

Communicating about Ratio and Rate Problems

You will need
- a calculator

▶ **GOAL** Explain your thinking when solving ratio and rate problems.

Problem 1

Stephen is making a scale model of the CN Tower.
The CN Tower is 553 m high and its base is 67 m across.

He decides to make his model 55.3 cm high.
How many cm wide should the base be?

Stephen's Explanation

I know the ratio of the height of the CN Tower to its base:

$$\frac{\text{tower height (m)}}{\text{base (m)}} = \frac{553}{67}$$

I'll make my model 55.3 cm high and use the same ratio:

$$\frac{\text{tower height (cm)}}{\text{base (cm)}} = \frac{55.3}{\quad}$$

So, $\frac{553}{67} = \frac{55.3}{6.7}$.

The base of my model should be 6.7 cm across.

Jasleen used the Communication Checklist to help
Stephen improve his explanation.

Jasleen's Questions

1. Why did you choose 55.3 cm to use for your model?

2. How did you calculate 6.7 cm?

Communication Checklist

- ☐ Did you identify the information given?
- ☐ Did you show each step of your calculation?
- ☐ Did you explain your thinking at each step?
- ☐ Did you check to see if your answer is reasonable?

Sofia earns $5/h while babysitting.
How much will she earn after babysitting for 10 h?

Sofia's Explanation

I know my rate is $5/h:

$$\frac{dollars}{hours} = \frac{\$5.00}{1\ h}$$

Using the same ratio, I can figure out what I will earn after babysitting for 10 h:

$$\frac{\$5.00}{1\ h} = \frac{\$50.00}{10\ h}$$

I know the scale factor is 10, because 1 x 10 = 10.
$5.00 x 10 = $50.00, so I will earn $50.00 in 10 h.

If I divide $50.00 by 10, I get $5.00.
If I divide $5.00 by 1, I get $5.00.
So, I know my answer is reasonable.

Reflecting

▶ What other questions could Jasleen have asked in Problem 1? Use the Communication Checklist for help.

_____

_____

▶ In Problem 2, did Sofia cover all the points in the Communication Checklist? Explain.

_____

_____

_____

Practising

Text page 52

3. Marlene ran 4 km in 30 min.
At this rate, can Marlene run 15 km in 120 min?
Use the Communication Checklist for help.

Communication Checklist

❑ Did you identify the information given?

❑ Did you show each step of your calculation?

❑ Did you explain your thinking at each step?

❑ Did you check to see if your answer is reasonable?

a) Write the information you are given to solve the problem.

_____

_____

b) You know that Marlene can run 4 km in 30 min. You want to know how many km she can run in 120 min.
Write this information as two equivalent rates with a missing term in the second rate.

$$\frac{\boxed{}}{\boxed{}} = \frac{\boxed{}}{\boxed{}}$$

c) Explain how you will find the scale factor to calculate the missing term.

_____

_____

d) Calculate the missing term in part b).

e) Can Marlene run 15 km in 120 min? How do you know?

_____

_____

f) Check to see if your calculation is reasonable. Show your work and explain your thinking.

7. A photocopier can make 1800 copies in 1 h. Rosa says that the photocopier can make 60 copies per min. Find the missing terms in the equivalent rates below.

$$\frac{1800 \text{ copies}}{1 \text{ h}} = \frac{1800 \text{ copies}}{\boxed{} \text{ min}} = \frac{\boxed{} \text{ copies}}{1 \text{ min}}$$

Is Rosa correct? How do you know?

_____

_____

_____

_____

_____

9. In a bulk store, the price of peanuts is $0.44 per 100 g. In a grocery store, the price of a 400 g bag of peanuts is $2.20.

a) Write the rates for each price in the table below.

Bulk store price	Grocery store price

b) Explain how you would compare the two rates.

_____

_____

c) Would you buy peanuts in the bulk store or the grocery store to get the most for your money? Explain your choice.

_____

_____

_____

2.5 Ratios as Percents

Text page 56

▶ **GOAL** Convert between ratios, percents, fractions, and decimals.

Follow these instructions to find the percent of shaded squares in each diagram.

A. Write the ratio of shaded squares to total squares in each diagram. The first diagram is done for you.

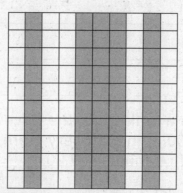

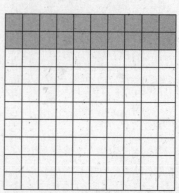

Ratio: 50 : 100

Ratio: ____ : ____

B. Write each ratio as a fraction.

Fraction: ▭/▭

Fraction: ▭/▭

Hint

Percent means per hundred.

C. Write each ratio as a percent.

Percent: _____

Percent: _____

D. Do both the ratios and percents compare a part to a whole? _____

Follow these instructions to find the percent of shaded squares in this diagram.

A. Write the ratio of shaded squares to total squares. This step is done for you.

2:5

B. Write the ratio as a fraction.

C. To find the percent, write an equivalent fraction with 100 as the bottom number.
Use a scale factor to find the missing term.

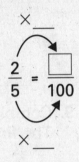

$$\frac{2}{5} = \frac{\square}{100}$$

D. Write the equivalent ratio as a percent. _____%

Reflecting

▶ How is writing a ratio as a percent like finding an equivalent ratio?

_____

_____

_____

_____

Practising

Text page 58 **7.** Complete the table by finding the missing equivalent fraction, ratio, decimal, or percent in each row.
Part a) is done for you.

	Fraction	Ratio	Decimal	Percent
a)	$\frac{1}{2}$	5:10	0.5	50%
b)		2:5		40%
c)	$\frac{21}{100}$		0.21	
d)		3:10	0.3	
e)	$\frac{1}{4}$			25%

9. Complete each calculation.
The first one is done for you.

a)

$$\overset{\times 5}{\frac{7}{20}} = \frac{35}{100} = 35\%$$
$$\underset{\times 5}{}$$

c)

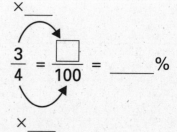

$$\frac{3}{4} = \frac{\square}{100} = \underline{\quad}\%$$

b)

$$\frac{23}{50} = \frac{\square}{100} = \underline{\quad}\%$$

11. The average rainfall for a region is 25 cm per day. On one day, 15 cm of rain fell.

    a) What is the ratio of rain that fell to the average rainfall?

        rain that fell:

        average rainfall:

$$\dfrac{\Box}{\Box}$$

    b) To find the percent, write an equivalent ratio that has 100 as its bottom number.

$$\dfrac{\Box}{\Box} = \dfrac{\Box}{100}$$

    c) What percent of the average rainfall has fallen?

        _____ %

13. Write each number as a percent. Then arrange them in order from greatest to least.

$$\dfrac{2}{8} = \dfrac{\Box}{100} = \text{_____} \%$$

$$0.22 = \dfrac{\Box}{100} = \text{_____} \%$$

$$\dfrac{3}{4} = \dfrac{\Box}{100} = \text{_____} \%$$

$$0.06 = \dfrac{\Box}{100} = \text{_____} \%$$

Order: _____, _____, _____, _____

END

Solving Percent Problems

You will need
• a calculator

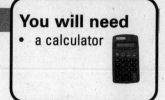

▶ **GOAL** Solve percent problems using decimals.

Percents are a special kind of ratio.
They are special because the denominator is always 100.

50% can be written as a fraction: $\frac{50}{100}$.

50% can also be written as 50:100 and as 0.50.

If there are 24 students in your class and half of them
are boys, then you could say that 50% of your
classmates are boys.

You might also write $\frac{12}{24}$.

When you divide 12 by 24 on your calculator, the answer
is 0.50. 0.50 is the same as 50%.

Problem 1

A CD is on sale for 25% off its original price.
The original price is $20.00. How much does the
CD cost?

Use these steps to solve the problem.

Step 1: Think 25% is the same as $\frac{25}{100}$ or 0.25.

Step 2: To find 25% of 20, multiply 0.25 by 20.

$$0.25 \times 20 = _____$$

Step 3: To find the new price of the CD, subtract your
answer in Step 2 from the original price of the CD.

$$20 - _____ = _____$$

So, the new price for the CD is _____.

Problem 2

Jack's job is to deliver 240 newspapers each day. On average, he delivers 168 newspapers before it gets dark. What percent of his job does he get done before dark?

Use these steps to solve the problem.

Step 1: Think of the problem as a ratio:

Jack delivers 168 out of 240 papers before dark.

Step 2: Write the ratio as a fraction.

Step 3: Divide 168 by 240 to find the decimal equivalent.

_____ ÷ _____ = _____

Step 4: Change the decimal to a percent.

$$\_\_\_\_\_ = \frac{\square}{10} = \frac{\square}{100} = \_\_\_\_\_\%$$

Reflecting

▶ Why is it helpful to know more than one way to write percents (for example, 50% can be $\frac{50}{100}$ or 0.50 or 50:100)?

_____

_____

_____

Practising

Text page 62 **7.** Write each percent as a decimal.
Then multiply to find the percent of each number.
The first one is done for you.

a) 50% of 20

$0.50 \times 20 = 10$

d) 12% of 50

_____ $\times$ _____ = _____

b) 75% of 24

_____ $\times$ 24 = _____

e) 15% of 200

_____ $\times$ _____ = _____

c) 20% of 45

_____ $\times$ 45 = _____

f) 44% of 250

_____ $\times$ _____ = _____

Text page 63 **9.** Out of a batch of 600 computers, 30 failed to pass
inspection due to faulty wiring.
What percent failed to pass inspection?

Think 30 out of 600 computers failed. The fraction is

$\dfrac{30}{600}$ ◄ numerator
◄ denominator

Find the decimal equivalent by dividing the
numerator by the denominator.

_____ $\div$ _____ = _____

Write the decimal as a percent. _____%

11. A dealer paid $6000 for a used car. The dealer wants to make a profit that is 25% of the price he paid for the car.

a) What profit does the dealer want to make?

Think 25% is the same as $\frac{25}{100}$ or 0.25.

0.25 × _____ = _____

The dealer wants to make $_____ in profit.

Hint

Add your answer in part a) to $6000.

b) How much should the dealer sell the car for?

_____ + _____ = _____

The dealer should sell the car for $_____.

13. A new chocolate bar is advertised as "20% MORE!" If the original chocolate bar is 50 g, what size is the new bar?

Think 20% is the same as $\frac{20}{100}$ or 0.20.

_____ × _____ = _____

Determine the size of the new chocolate bar in grams.

_____ + _____ = _____

The new chocolate bar is _____g.

Decimal Multiplication

▶ **GOAL** Multiply a decimal by a decimal.

Problem 1

Mark is asked to multiply 0.4 × 0.7.

Use these steps to help Mark model 0.4 × 0.7 on a 10-by-10 grid.

Hint

Each square on the grid represents 0.01. Each row represents 0.1. Each column represents 0.1. The grid represents 1 whole.

Step 1: Mark shaded columns on the grid. ▶

What decimal is represented by the shaded columns?

_____

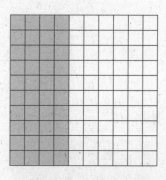

Step 2: Then Mark shaded rows of the grid in a darker colour to represent 0.7. ▶

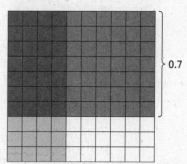

0.7

Step 3: Count the number of squares in the overlap area (the darkest area).

number of overlap squares: _____

Step 4: The overlap area represents the product of 0.4 × 0.7.

Hint

Remember, there are 100 squares on the grid.

What fraction of the entire grid do the overlap squares represent?

$\dfrac{\boxed{}}{100}$

Write the fraction as a decimal. _____

So, 0.4 × 0.7 = _____

Tara can ride her bike at an average speed of 14.8 km/h. At this speed, how far will she travel in 2.25 h?

To understand the problem, remember that distance = speed × time.
So, you have to multiply 14.8 km/h (speed) by 2.25 h (time) to find the distance Tara will travel.

Use these steps to estimate and calculate the answer.

Step 1: Round the numbers to whole numbers.

   2.25 h is about _____ h.

   14.8 km/h is about _____ km/h.

Step 2: Multiply the rounded numbers to estimate.

   _____ × _____ = _____

   So, she will travel about _____ km in 2.25 h.

Step 3: Use a calculator to calculate the product and check your estimate.

   2.25 × 14.8 = _____

Step 4: Compare your answer in Step 3 with your estimate in Step 2. Was your estimate reasonable? _____

Reflecting

▶ Why is the product of 0.4 × 0.7 less than both the numbers?

_____

_____

_____

_____

Practising

Text page 66 **8.** Use a 10-by-10 grid to model the answer.
Then complete the statements.

a) 0.2 × 0.6 = _____

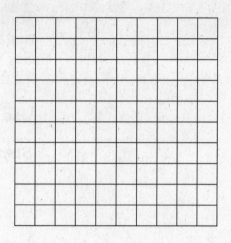

b) 0.8 × 0.7 = _____

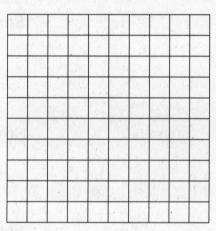

c) 0.2 × 0.7 = _____

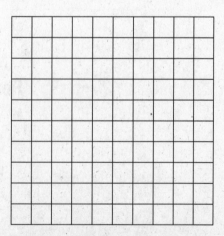

d) Why is it not practical to model the product of
3.7 × 11.2 using grids?

_____

_____

9. Round the numbers to estimate each product.
Then check your estimates with a calculator.

	Multiplication sentence	Estimate	Circle the answer closest to your estimate			Check with a calculator
a)	$0.9 \times 6.1 =$	$1 \times 6 =$ ___	5.49	54.9	549	
b)	$0.8 \times 1.3 =$		0.104	1.04	10.4	
c)	$2.6 \times 10.1 =$		2.626	26.26	262.6	

Text page 67

12. A car travels for 3.5 hours at an average speed of 92.5 km/h. How far did the car travel?

a) Round the numbers to estimate the distance travelled.

3.5 is about _____.

92.5 is about _____.

Hint
Remember, distance = speed $\times$ time.

b) Multiply the rounded numbers.

_____ $\times$ _____ = _____

c) Calculate the distance travelled. Use a calculator.

Distance: $3.5 \times 92.5 =$ _____

So, the car travelled _____ km.

d) Compare your answer in part c) with your estimate in part b).

Was your estimate reasonable? _____

END

Decimal Division

You will need
• coloured pencils

▶ **GOAL** Divide decimals.

Problem 1

You have 2 L of orange juice and you want to divide it among your friends. How many glasses can you fill?

Use these steps to solve the problem using grids.

Step 1: One glass holds 0.35 L. Each grid represents 1 L. Each square represents 0.01 L. Shade 0.35 on the grid. This step is done for you.

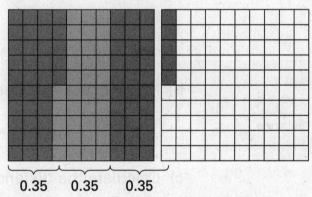

0.35   0.35   0.35

Step 2: Use alternating colours to shade 0.35 as many times as you can. The next two 0.35 sections are shaded for you.

Step 3: How many glasses can be filled from the 2 L container? _____

How much juice is left over? _____

Step 4: Write the solution as a division sentence.

_____ ÷ _____ = _____ remainder _____

How would you find the answer if the container held only 1.5 L of juice?

Hint

Think about how much of the two grids you would need.

_____

Problem 2

Katya has 2.4 m of string. She wants to divide it into equal pieces. How many pieces will there be if each piece is 0.4 m long?

Use these steps to solve the problem using equivalent ratios.

Step 1: Write the division sentence as a ratio. This step is done for you.

$$2.4 \div 0.4 = \frac{2.4}{0.4}$$

Hint

Multiply by 10 or 100 to find an equivalent ratio that has only whole numbers.

Step 2: Find an equivalent ratio that has only whole numbers. Then divide.

$$= \frac{2.4 \times \boxed{}}{0.4 \times \boxed{}}$$

$$= \frac{\boxed{}}{\boxed{}}$$

$$= \underline{\qquad} \div \underline{\qquad}$$

$$= \underline{\qquad}$$

Step 3: How many pieces of string will there be? _____

Reflecting

▶ In Step 2 above, why do you need to multiply both numbers in the ratio by the same number?

_____

_____

▶ Why does $\frac{24}{4}$ give the same answer as $\frac{2.4}{0.4}$?

_____

_____

Practising

Text page 70

5. Use the 10-by-10 grids to model, and then calculate.

a) 2.7 ÷ 0.9 = _____

> **Hint**
> Make sure you only use 2.7 of the three grids.

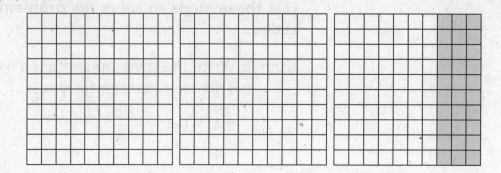

> **Hint**
> Make sure you only use 3.6 of the four grids.

b) 3.6 ÷ 0.18 = _____

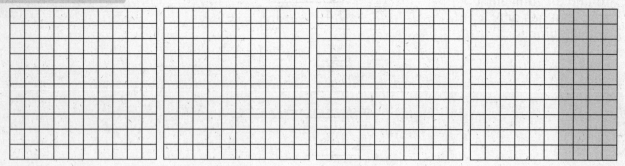

9. Divide. The first one is partly done for you.

a) $10.2 \div 1.5 = \dfrac{10.2}{1.5}$

$$= \dfrac{10.2 \times 10}{1.5 \times 10}$$

$$= \dfrac{\boxed{}}{\boxed{}} \qquad \overline{\smash{\big)}}$$

b) $14.4 \div 0.12 = \dfrac{\boxed{}}{\boxed{}}$

> **Hint**
> Remember, you must multiply each number by the same number.

$$= \dfrac{\boxed{} \times \boxed{}}{\boxed{} \times \boxed{}}$$

$$= \dfrac{\boxed{}}{\boxed{}} \qquad \overline{\smash{\big)}}$$

13. Nathan has 11.4 m of rope. He wants to divide it into equal pieces. How many pieces will there be if the pieces are 0.8 m long?

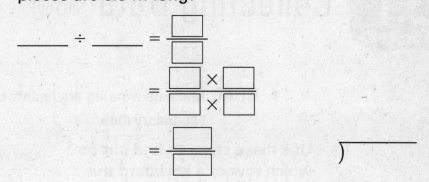

$$\underline{\hspace{1cm}} \div \underline{\hspace{1cm}} = \frac{\boxed{}}{\boxed{}}$$

$$= \frac{\boxed{} \times \boxed{}}{\boxed{} \times \boxed{}}$$

$$= \frac{\boxed{}}{\boxed{}}$$

$$\overline{\smash{)}\hspace{2cm}}$$

14. How long will it take to walk 10 km at 4.5 km/h? Round your answer to the nearest hour.

$$\underline{\hspace{1cm}} \div \underline{\hspace{1cm}} = \frac{\boxed{}}{\boxed{}}$$

$$= \frac{\boxed{} \times \boxed{}}{\boxed{} \times \boxed{}}$$

$$= \frac{\boxed{}}{\boxed{}}$$

$$\overline{\smash{)}\hspace{2cm}}$$

Collecting Data

You will need
• a calculator

▶ **GOAL** Make convincing arguments based on primary and secondary data.

Use these steps to find out on which row of a keyboard the most frequently used letters are found.

Step 1: On the table, mark one tally beside a letter each time the letter occurs in the paragraph below. The crossed out words have been tallied for you.

Step 2: Count the number of tallies for each letter. Write the number in the Frequency column of the table.

~~Traditional Triangular Nim~~

~~Set up your rows with one counter in the first row, two counters in the second row, three in the third row, four in the fourth row, and five in the fifth row.~~

Players take any number of counters from one row on a turn. The player who takes the last counter loses the game.

Letter	Tally	Frequency
A	IIII	
B		
C	III	
D	IIII	
E	IIII IIII III	
F	IIII I	
G	I	
H	IIII IIII	
I	IIII IIII IIII	
J		
K		
L	II	
M	I	
N	IIII IIII III	
O	IIII IIII IIII	
P	I	
Q		
R	IIII IIII IIII II	
S	IIII	
T	IIII IIII IIII III	
U	IIII II	
V	I	
W	IIII III	
X		
Y	I	
Z		
Total		**235**

This table shows **secondary data** for how frequently letters are used in the English language.

Letters in each row	Total frequency	Percent
QWERTYUIOP (top row)	540	$\frac{540}{1000} = 54\%$
ASDFGHJKL (middle row)	299	$\frac{299}{1000} = 30\%$
ZXCVBNM (bottom row)	161	$\frac{161}{1000} = 16\%$
Total	**1000**	**100%**

Step 3: Complete this table using the **primary data** from your tally chart.

Letters in each row	Total frequency	Percent
QWERTYUIOP (top row)		$\frac{}{235} = $ _____ %
ASDFGHJKL (middle row)		$\frac{}{235} = $ _____ %
ZXCVBNM (bottom row)		$\frac{}{235} = $ _____ %
Total	**235**	$\frac{235}{235} = 100\%$

Reflecting

▶ What do you notice when you compare your percents to the percents in the secondary data?

_____

_____

_____

_____

_____

END

Avoiding Bias in Data Collection

▶ **GOAL** Understand how to avoid bias in data collection.

Read this example to see how bias could occur in data collection.

Data is collected about the favourite television programs of 12-year-olds. Homes are phoned between 2:00 p.m. and 3:00 p.m. on a school day to gather data. Would these results be biased?

Solution

Yes. These results would be biased because most 12-year-olds are still at school between 2:00 p.m. and 3:00 p.m. So students who are at school would not be surveyed.

MATH TERM

survey
asking a group of people a question to collect data about the group

Hint
Results are biased when a distinct part of the group you are trying to find data for is not surveyed.

A. Explain why each **survey** below is biased. The first one is done for you.

Survey question	Survey idea	Biased because:
What time should school start in the morning?	The students of Grade 7 home room 7GF are asked.	Teachers, administrators, and other grades are not surveyed.
	The students who came late on Tuesday are asked.	

B. Think of a way to conduct each survey below to avoid bias. Then explain why the survey avoids bias. The first one is done for you.

Survey question	Survey idea that avoids bias	Avoids bias because:
What are the most popular television shows for families in your area?	A company conducts a telephone survey by calling every 100th name in the phone book between 6:00 p.m. and 8:00 p.m. Calls are made again if no one answers.	• By calling every 100th name in the phone book, all groups of the population have a chance to answer. • Most families are home between 6:00 p.m. and 8:00 p.m.
What do students think is the best day of the week?		

Hint

To avoid bias, try to think of a way to survey people where you will not miss a distinct part of the group.

Reflecting

▶ When conducting a survey, why should you try to avoid bias?

_____

_____

_____

_____

Practising

Text page 86

8. Explain why the survey is biased.
Discuss your answer with at least one other person.

	Survey question	Survey idea	Biased because:
a)	Should hunting be banned in Ontario?	100 people are asked at a shopping mall in Toronto.	
b)	How much time do you spend on a computer each week?	The survey is conducted over the Internet.	
c)	Which sport is more popular in Ontario— hockey or soccer?	People are asked as they enter a hockey arena.	

9. Think of a way to conduct each survey to avoid bias.
Then explain why the survey avoids bias.
Discuss your answers with at least one other person.

	Survey question	Survey idea that avoids bias	Avoids bias because:
a)	What is the favourite type of music for people in your community?		
b)	Should the school build a rock climbing wall or a bigger library?		

END

Using a Database

▶ **GOAL** Use a database to sort and locate data.

The **database** below contains data about eight
students' pets.

field
a category (usually a column)

record
all the data
about one
item (usually
a row)

Student number	Last name	First name	Type of pet	Age of pet	Weight of pet (kg)	Name of pet
123	Brown	Debal	dog	2.5	27.3	Buddy
345	Dee	Alex	cat	1	8.2	Fifi
352	Forester	Rachel	cat	0.5	2.2	Beau
876	Khan	Fazil	fish	2	0.1	Noire
901	Miller	Tom	rabbit	2.5	9.4	Sammy
442	Shepherd	John	white rat	0.5	0.3	Boris
171	Taylor	Robin	gerbil	1.5	0.5	Zippy
11	Unrau	Sarah	dog	8	17	Spot

Record: |◄ ◄   1 ► ►| ►* of 8

entry
a single piece of data

A. Write the fields shown in this database.
The first two fields are written for you.

Student number, Last name, _____

_____

B. This database is sorted by the "Last name" field in
alphabetical order.

Which "Last name" will appear first if the database is
sorted by

• Student number, from greatest to least? _____

• Weight of pet, from least to greatest? _____

C. Which field would you sort the database by to find
the oldest pet?

_____

Reflecting

▶ Who might keep information about you in a database? How would they use this information?

_____

_____

Practising

Text page 91

7. This database lists information about different countries in the world.

Country	Continent	Area (square km)	Population (millions)
Australia	Australia	7 687 000	19.4
Brazil	S. America	8 512 000	174.4
Canada	N. America	9 976 000	31.6
China	Asia	9 597 000	1273.0
Congo	Africa	342 000	53.6
Iceland	Europe	103 000	0.3
Jamaica	N. America	11 000	2.7
Niger	Africa	1 267 000	10.4
Singapore	Asia	620	4.3
U.S.A.	N. America	9 373 000	278.1

Record: |◄ ◄ 4 ► ►| ►* of 10

a) Circle the field that this database is sorted by.

   Country   Continent   Area   Population

b) If you sorted this database by Population from greatest to least, which country would appear first?

   _____

c) If Iceland appeared last, which field would you have used to sort?

   _____

d) If you sorted this database by Area from least to greatest, what would be the first three countries?

   first: _____

   second: _____

   third: _____

END

Using a Spreadsheet

▶ **GOAL** Understand the difference between a spreadsheet and a database.

Samir is determining the cost of school supplies. He uses this spreadsheet.

cell
the intersection of a column and a row, where a single piece of data is stored; for example, cell B3 is the intersection of column B and row 3

	A	B	C	D
1	**Item**	**Unit Price**	**Quantity**	**Total**
2				
3	Pencils	$0.50	4	$2.00
4	Graph paper	$1.50	2	$3.00
5	Ruler	$1.00	1	$1.00
6	Notebook	$4.00	1	$4.00
7				
8			Subtotal	$10.00
9			PST (8%)	$0.08
10			GST (7%)	$0.07
11			Total	$10.15

A. Column A shows the names of each item that Samir bought.

What information is in column B?

_____

What information is in column D?

_____

_____

B. Write the data found in each cell. The first one is done for you.

A3: _____Pencils_____

B6: _____

C5: _____

D11: _____

C. The total cost of the pencils is found in cell D3. This cost was calculated by multiplying the unit price ($0.50) by the quantity (4). A spreadsheet can make this calculation for you.

The total cost of graph paper ($3.00) is found in cell D4. How was this cost calculated?

_____

_____

D. What cells were used to calculate the subtotal in cell D8?

_____

How was the subtotal calculated?

_____

_____

E. The total cost, including tax, is found in cell D11. How was this cost calculated?

_____

_____

F. Samir decides to buy 4 rulers. Shade all the cells in the spreadsheet that change. Write the changes beside the old numbers in the spreadsheet.

Reflecting

▶ Spreadsheets look a lot like databases. What can you do on a spreadsheet that you cannot do on a database?

_____

_____

END

Frequency Tables and Stem-and-Leaf Plots

▶ **GOAL** Organize data using frequency tables and stem-and-leaf plots.

In a standing long jump competition, students jumped the following distances (in cm):

187	205	221	186	185	212	222	215	198
200	205	207	193	186	172	208	223	175

MATH TERM

frequency table
shows the count of each item in an interval

Tonya decides to create a frequency table to sort the data.

Step 1: Tonya looks at the range of the data.

Range: _____ − _____ = _____
        (longest jump)   (shortest jump)

She knows that 51 divided by 10 is just over 5. If she chooses an **interval** of 10, she will have just over 5 intervals.

MATH TERM

interval
the space between two numbers; 0–9 is an interval of 10, this interval includes 0, 9, and all the numbers between 0 and 9

Interval (cm)	Tally	Frequency
170–179		
180–189	I	
190–199		
200–209	I	
210–219		
220–229		

Hint

In the table, Tonya starts at 170 (instead of the shortest jump 172) because the intervals are easier to work with.

She ends at 229 because the longest jump is between 220 and 229.

Step 2: In the frequency table, tally the number of distances that occur in each interval. The first two numbers, 187 and 205, are done for you.

Step 3: Count the number of tallies for each interval. Write the number in the Frequency column of the table.

stem-and-leaf plot organizes data based on place values

Darren decides to create a stem-and-leaf plot to sort the data.

In this stem-and-leaf plot, the hundreds and tens digits are the stem. The ones digit is the leaf.
For example, for the number 187, the stem is 18 and the leaf is 7.

Step 1: Use the standing long jump data on page 70 to complete the stem-and-leaf plot below. The first two numbers from the data, 187 and 205, are done for you.

Hint

The first stem is 17 because the shortest jump is 172.

The last stem is 22 because the longest jump is 223.

Long Jump Distances (cm)	
Stem	Leaf
17	
18	7
19	
20	5
21	
22	

Step 2: Order the numbers in the Leaf column from least to greatest. The first two rows are done for you.

Long Jump Distances (cm)	
Stem	Leaf
17	2 5
18	5 6 6 7
19	
20	
21	
22	

TURN

Reflecting

▶ How is a stem-and-leaf plot like a frequency table?

_____

_____

▶ (Circle) the method in which individual data values are lost.

frequency table   stem-and-leaf plot

▶ What method do you find most useful for comparing data? Explain.

_____

_____

_____

_____

Practising

Text page 100

5. Organize each set of data in a frequency table.

a) number of words in short stories

| 120 | 173 | 287 | 599 | 183 | 298 | 376 | 452 |

Range: _____ – _____ = _____
     (greatest)   (least)

Interval of _____

Interval	Tally	Frequency
100 – _____		
200 – 299		
300 – _____		
400 – _____		
_____ – 599		

b) heights of plants (cm)

120	387	428	127	287	125	332	487

Range: _____ – _____ = _____
      (greatest)   (least)

Interval of _____

Interval	Tally	Total count
–		
–		
–		
–		

Text page 101 **10.** These are Rosa's bowling scores for a season.

132	118	122	106	94	94	112	118	104	120
108	104	96	122	130	116	104	118	106	124

a) What is the range of the data?

_____ – _____ = _____
(greatest)   (least)

b) Display Rosa's scores in a stem-and-leaf plot.
Order the numbers in the Leaf column from least
to greatest.

Bowling Scores	
Stem	**Leaf**
9	
10	
13	

Lesson 3.5: Frequency Tables and Stem-and-Leaf Plots **73**

END

Mean, Median, and Mode

▶ **GOAL** Describe data using the mean, median, and mode.

Problem 1

Bella's math team had the following contest scores:

81	73	74	86	63	79	90	81

Follow these instructions to find the mean, median, and mode of the contest scores.

MATH TERM

mean
the sum of a set of numbers divided by the number of numbers in the set

A. Calculate the **mean** of the scores. Fill in the missing scores below. Then use a calculator.

$$\frac{81 + 73 + + + + + 90 + 81}{8} = \frac{\boxed{}}{8}$$

$$= \underline{}$$

MATH TERM

median
the number that is in the middle of a set of numbers

B. Find the **median** of the scores.

Step 1: Order the scores from least to greatest. This step is done for you.

$$63 \quad 73 \quad 74 \quad 79 \;\Big|\; 81 \quad 81 \quad 86 \quad 90$$

Step 2: Circle the scores that are on either side of the middle line.

Step 3: Find the mean of the two middle numbers. This calculation gives you the median.

$$\frac{\boxed{} + \boxed{}}{2} = \frac{\boxed{}}{2}$$

$$= \underline{}$$

MATH TERM

mode
the number that occurs most often in a set of numbers

C. Write the **mode** of the scores. _____

Problem 2

Chocolate bars are on sale at different stores for the prices shown in this stem-and-leaf plot.

Follow these instructions to find the mean, median, and mode in this stem-and-leaf plot.

Chocolate Bar Prices (¢)	
Stem	Leaf
7	7
8	5 5 7 8 9
9	3 3 3
10	0 5

A. Complete the list of prices represented in the stem-and-leaf plot.

77¢

85¢, 85¢, 87¢, 88¢, 89¢

93¢, 93¢, _____¢

100¢, _____¢

B. Use your calculator to determine the mean of the prices.

$$\frac{77 + 85 + 85 + \quad + \quad + \quad + \quad + \quad + \quad + 105}{11} = \frac{\boxed{}}{11}$$

$$= \underline{\quad\quad}¢$$

C. Use these steps to find the median price.

Step 1: How many prices are there? _____

Step 2: Circle the leaf on the stem-and-leaf plot that has an equal number of leaves before and after it.

Step 3: Write the median. _____¢

D. Find the mode of the prices. _____¢

Hint

When there is an odd number of numbers, the median is the middle number.

Reflecting

▶ Suppose the price of the 77¢ chocolate bar decreased to 66¢. Why does the median stay the same?

_____

Practising

4. Calculate the mean for each set of numbers.

a)

23	52	40	23	56	96

Mean: $\dfrac{\boxed{}}{\boxed{}} = \dfrac{\boxed{}}{\boxed{}}$

$= \underline{}$

b)

6.2	7.4	8.3	5.7	4.3

Mean: $\dfrac{\boxed{}}{\boxed{}} = \dfrac{\boxed{}}{\boxed{}}$

$= \underline{}$

5. Find the median for each set of numbers.
Order the data from least to greatest first.

a)

4	8	2	9	3	3	0

_____, _____, _____, _____, _____, _____, _____

The median is _____.

b)

32	88	13	54	84

_____, _____, _____, _____, _____

The median is _____.

6. Find the mode for this set of numbers.

7	6	9	9	8	6	6	4

The mode is _____.

9. This stem-and-leaf plot shows Martin's bowling scores for the season.

Bowling Scores	
Stem	**Leaf**
10	4 6
11	4
12	2 8 9
13	2 2

a) Calculate his mean bowling score.

$$\frac{\boxed{}}{\boxed{}} = \frac{\boxed{}}{\boxed{}}$$

$$= \underline{}$$

Hint

Check to see that there are an equal number of scores before and after the circled numbers.

b) To find the median score, ⟨circle⟩ the two middle numbers in the stem-and-leaf plot.

Calculate the mean of the middle numbers. Remember to include the stem.

$$\frac{\boxed{} + \boxed{}}{2} = \frac{\boxed{}}{2}$$

$$= \underline{}$$

Martin's median score is _____.

c) Find the mode of Martin's bowling scores.

_____

Communicating about Graphs

▶ **GOAL** Make convincing arguments based on trends.

Janet created a graph and wrote a report to show how much milk her school should order each week. Read her report and answer the questions below.

Janet's Report

The graph shows more people drink chocolate milk at the beginning of the week. We should order 100 white milks each week since there are five days in a week and the most sold in a day is 24. We should order 150 chocolate milks each week since the most sold in a day is 33.

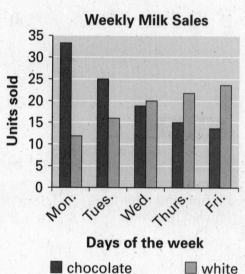

Weekly Milk Sales

Units sold / Days of the week

■ chocolate   ■ white

A. Describe any important details that Janet did not include.

_____

_____

_____

_____

B. Are Janet's conclusions reasonable? Explain.

_____

_____

_____

Reflecting

▶ What parts of the Communication Checklist did Janet cover well? Explain.

_____

_____

_____

Practising

5. Robert surveyed all 48 family members at his family reunion about their favourite kinds of pie. Then he drew this graph.

Write a report about the types of pie Robert's family should buy for next year's reunion. Use the Communication Checklist for help.

Favourite Kinds of Pie

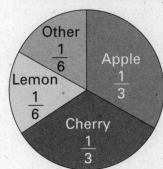

a) How does the graph help you decide how many of each type of pie Robert's family should buy?

_____

_____

b) What advice could Robert give about the pies to serve at the next family reunion?

_____

_____

_____

Exploring Number Patterns

You will need
• a calculator

▶ **GOAL** Identify and describe number patterns.

MATH TERM

Pascal's triangle
a triangle made up of whole numbers with many different number patterns in it

Find patterns in **Pascal's triangle** by answering the questions below.

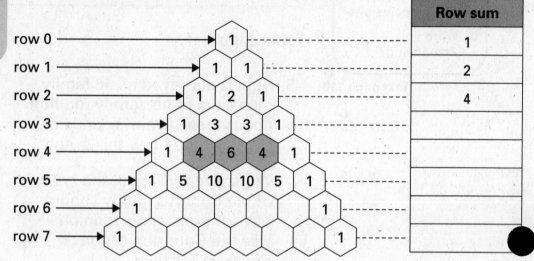

	Row sum
row 0 → 1	1
row 1 → 1 1	2
row 2 → 1 2 1	4
row 3 → 1 3 3 1	
row 4 → 1 4 6 4 1	
row 5 → 1 5 10 10 5 1	
row 6 → 1 ... 1	
row 7 → 1 ... 1	

A. What is the relationship between the numbers in row 3 and the shaded numbers in row 4?

_____

_____

Hint

Look at two numbers in row 3 that touch one shaded number in row 4.

B. Does the pattern work for other rows? _____ Give an example.

_____

C. Use the pattern to complete the rest of the triangle.

D. Add the numbers in each horizontal row and enter the sums in the table.

E. Look for a pattern in the sums of the rows. Describe the pattern.

_____

F. Describe the "hockey stick" pattern shaded in the triangle below. There are two "hockey sticks" shaded.

_____

_____

G. Use the triangle on page 80 to complete the triangle below. Shade one more "hockey stick."

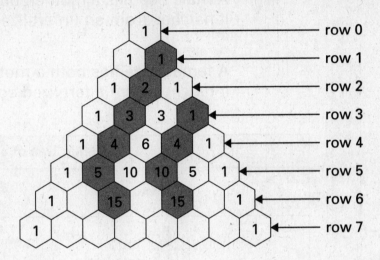

row 0
row 1
row 2
row 3
row 4
row 5
row 6
row 7

Hint

Look at the numbers on the diagonals and the row numbers.

Reflecting

▶ Which four numbers in row 7 of Pascal's triangle could you fill in without knowing the numbers in row 6? Explain how you would do this.

_____

_____

_____

_____

▶ Which numbers could you not fill in unless you knew the numbers in row 6? Explain why not.

_____

_____

_____

_____

END

Applying Pattern Rules

You will need
• a calculator

▶ **GOAL** Recognize patterns, and use rules to extend patterns.

A male bee has a mother, but no father.
It hatches from an unfertilized egg.

A female bee has both a mother and a father.
It hatches from a fertilized egg.

Family Tree of a Bee		
Pictures	**Number of bees**	**Add to get next number**
male bee	1	+0
female bee (mother of male bee)	1	+1
(father and mother of female bee)	2	+1
	3	+2
	5	

Use these steps to find the number pattern in the family tree of a bee.

Step 1: Add one more row of pictures to the family tree.

Step 2: Complete the second column of the table.

Step 3: Complete the third column of the table.

Step 4: Look at any number in the third column. Then look at the number in the second column in the row above. What do you notice?

_____

Is this the same for every number in the table?

_____

Step 5: Describe the pattern rule in this number sequence.

_____

_____

Reflecting

▶ Could you find the number of bees in the 12th row without knowing the number of bees in the 10th and 11th rows? Explain your answer.

_____

_____

_____

_____

Connect Your Work

Describe the pattern rule for each sequence below.
Write the next four numbers in the sequence.

1. Pattern rule: _____

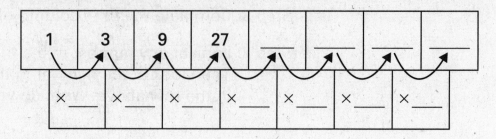

2. Pattern rule: _____

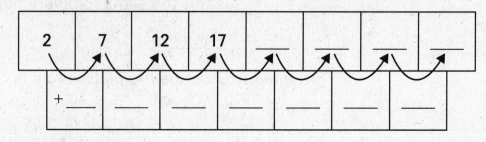

Practising

Text page 126

5. Describe the pattern rule for each sequence below.
Write the next three numbers.

a) Pattern rule: _____

4	8	16	32	_____	_____	_____

b) Pattern rule: _____

4	8	12	16	20	_____	_____	_____

6. Describe the pattern rule for the sequence below. Write the next three numbers.

Pattern rule: _____

_____

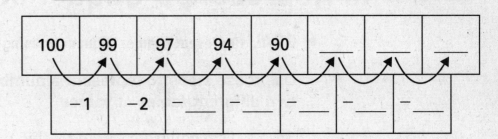

8. a) Describe the pattern rule for the sequence below.

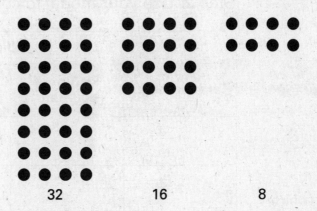

32   16   8

Pattern rule: _____

> **Hint**
>
> The last number can be a fraction or a decimal.

b) Write the next four numbers in the sequence.

32, 16, 8, _____, _____, _____, _____

10. The pattern rule for a number sequence is: "Start at 2, triple each number, and add 1."

Write the next four numbers.

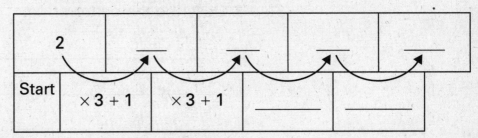

Using a Table of Values to Represent a Sequence

You will need
- a calculator
- counters

▶ **GOAL** Represent number sequences using tables.

Use these steps to represent a number sequence with two different tables of values.

Step 1: Use counters to model this pattern: 3, 5, 7, 9, ...

Step 2: Use your model to complete Table 1.

Table 1			
Term number	Picture of counters	Term value (number of counters)	Add to get next term value
1	○ ○○	3	+2
2	○○ ○○○	5	+2
3	○○○ ○○○○	7	
4	○○○○ ○○○○○	9	
5			
6			

Step 3: Describe the pattern rule.

_____

_____

Step 4: Describe the pattern rule used in Table 2 below.

_____

Table 2		
Term number	**Pattern rule**	**Term value**
1	$1 \times 2 + 1$	3
2	$2 \times 2 + 1$	5
3	$3 \times 2 + 1$	
4		
5		
6		

Step 5: Complete Table 2 using the pattern rule.

Reflecting

▶ How are Table 1 and Table 2 the same?

_____

_____

▶ How are Table 1 and Table 2 different?

_____

_____

▶ If you wanted to find the value of the 35th term, which of the two pattern rules would you use? Explain.

_____

_____

_____

_____

_____

Practising

Text page 131 5. Peter and Heidi are looking at this table of values. ▶

Term number	Term value
1	3
2	6
3	9
4	12
5	15

a) Peter says the pattern rule is "Start with 3 and add 3 each time."

Complete the table below using Peter's rule.

Term number	Peter's rule	Term value
1	3	3
2	3 + 3	
3	6 + 3	
4		
5		

b) Heidi says the pattern rule is "Multiply the term number by 3."

Complete the table below using Heidi's rule.

Term number	Heidi's rule	Term value
1	1 × 3	3
2	2 × 3	
3		
4		
5		

c) Were they both right? _____

Explain why. _____

_____

_____

7. Complete the table of values for this sequence. Include pictures.

Term number	Picture	Term value
1	ooooo	5
2	ooooo oooo	9
3		13
4		17
5		
6		
7		

10. Asha has only 38 toothpicks.
If she uses as many of these toothpicks as she can, what is the greatest figure number she can build in this pattern?

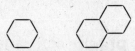

Figure 1   Figure 2

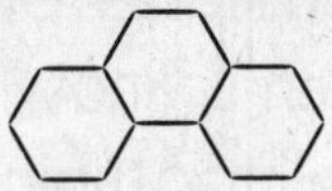

Figure 3

Term number (Figure number)	Pattern rule	Term value (number of toothpicks)
1	Start at 6	6
2	+ _____	11
3	+ _____	
4	+ _____	
5	+ _____	
6	+ _____	
7	+ _____	
8	+ _____	

Look for a pattern rule.
Then complete the table of values to find the answer.

Greatest figure number if she has only 38 toothpicks:

_____

END

Solve Problems Using a Table of Values

You will need
- a calculator
- toothpicks

▶ **GOAL** Solve patterning problems using tables.

Problem 1

Tynessa wants to make Figure 6 of the pattern below. How many toothpicks does she need?

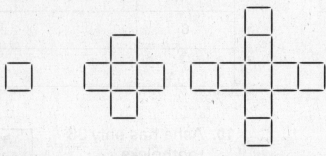

Figure 1   Figure 2   Figure 3

1 Understand the Problem

I have to find out how many toothpicks are in Figure 6. I know how many toothpicks are in Figures 1, 2, and 3 by looking at the picture above.

2 Make a Plan

I will make a table of values and look for a number pattern. When I know the number pattern, I can find the number of toothpicks needed to make Figure 6.

3 **Carry Out the Plan**

Step 1: Complete the table of values for Figure 3.

Figure number	Picture	Total number of toothpicks
1	□	4
2		16
3		
4		
5		
6		

Step 2: Look for a pattern rule in the table.
Describe the pattern rule.

_____

Step 3: Complete the table using your rule.

4 **Look Back**

To check your answer, draw Figure 6 on a scrap piece of paper.

Hint

Once you know the pattern rule, you do not have to draw any more models of the figures.

Reflecting

▶ How did making a table of values help you decide how many toothpicks you would need to make Figure 6?

_____

_____

_____

Practising

Text page 137 **10.** Heather wants to make 250 origami cranes for a school display.

a) Look for a pattern in the table of values. Then complete the table to determine the total number of cranes made by the end of each day.

Day	Number of cranes made that day	Total number of cranes made
1	12	12
2	15	15 + 12 = 27
3	18	18 + 27 = _____
4	21	21 + _____ = _____
5		_____ + _____ = _____
6		_____ + _____ = _____
7		_____ + _____ = _____
8		_____ + _____ = _____
9		_____ + _____ = _____
10		_____ + _____ = _____

> **Hint**
>
> In the third column, add the number of cranes made that day to the total number of cranes made in the previous days.

b) If she continues this pattern for 10 days, will she reach her goal of 250 cranes? _____

12. The students challenge the teachers to a basketball shootout. The money from the ticket sales will go to charity.

- On the first day, 16 tickets are sold.
- On the following days, the number of tickets sold increases by 3 tickets per day.
- Each ticket costs $2.00.
- The goal is to raise $175.00.

a) Complete the table to show the number of tickets sold each day.

Day	Number of tickets sold that day	Total number of tickets sold	Cost of tickets sold ($2.00/ticket)
1	16	16	$32.00
2	19	19 + 16 = 35	
3		_____ + 35 = _____	
4		_____ + _____ = _____	
5		_____ + _____ = _____	

b) What is the total number of tickets sold by the end of day 5?

_____

c) On which day will the students reach their goal of $175.00?

_____

END

4.5

Using a Scatter Plot to Represent a Sequence

▶ **GOAL** Graph number sequences using scatter plots.

Follow these instructions to use a scatter plot to represent this number sequence:

4, 6, 8, 10, …

A. Extend the number pattern to complete the table of values. ▶

Term number	Term value
1	4
2	6
3	8
4	10
5	
6	

MATH TERM

coordinates
an ordered pair, used to describe a location on a grid with an x-axis and a y-axis; for example, the coordinates (2, 3) describe this location:

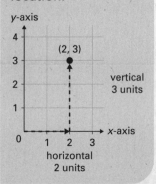

B. The term number and term value in the same row of the table of values are **coordinates** for a point on the scatter plot.

Complete the scatter plot using the data from the table of values above.

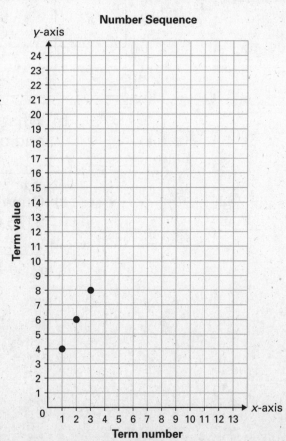

Number Sequence

C. Use these steps to predict the term value for term number 12.

Step 1: Line up your ruler with the points on your scatter plot.

Step 2: Draw a dashed line through the points up to the top of the scatter plot.

Step 3: Look for term number 12 on the *x*-axis. Go straight up to the dashed line.

Step 4: Then go straight left to the term value on the *y*-axis.

Term value for term number 12: _____

D. Use these steps to predict the term number that has a term value of 18.

Step 1: Look for term value 18 on the *y*-axis. Go straight across to the dotted line.

Step 2: Then go straight down to find the term number on the *x*-axis.

Term number of term value 18: _____

E. Extend the pattern in the table of values in the margin.

What is the term value for term number 12? _____

Did you get the same answer as you did in Part C?

_____

Term number	Term value
1	4
2	6
3	8
4	10
5	
6	
7	
8	
9	
10	
11	
12	

Reflecting

▶ When making predictions about a sequence, why is using a scatter plot not as accurate as using a pattern rule from a table of values?

_____

_____

_____

TURN ▶

Practising

Text page 140

5. Mohammed designed a fence that has 5 logs in the first section. It has 4 logs in each section after that.

section 1   section 2   section 3

a) How many logs are in the first two sections? _____

b) Complete the table of values below.

c) Create a scatter plot using the table of values.

Number of sections	Number of logs
1	5
2	
3	
4	
5	

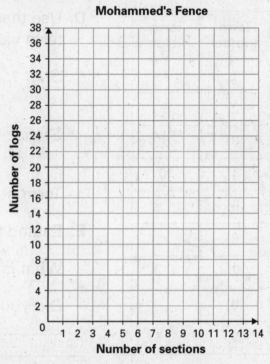

d) Use the scatter plot to predict how many logs Mohammed will need for a fence that is 8 sections long.

Number of logs: _____

7. a) Create a scatter plot using the table of values below.

b) Use your scatter plot to find the term value for term number 20.

Term value for term number 20: _____

Term number	Term value
4	8
6	10
8	12
10	14
12	16

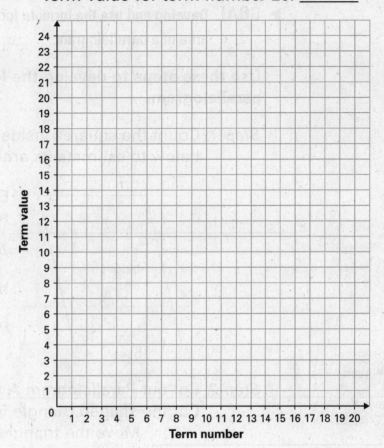

Lesson 4.5: Using a Scatter Plot to Represent a Sequence **97**

END

Area of a Parallelogram

You will need
- Cutout Page 5.1
- scissors
- tape
- a ruler

- a calculator

▶ **GOAL** Develop and use the formula for the area of a parallelogram.

Use these steps to develop the formula for the area of a parallelogram.

Step 1: Count the squares inside the parallelogram below to estimate its area.

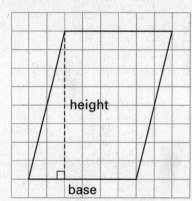

Each square on the grid represents 1 cm².

Area = about _____ cm²

base = _____ cm

height = _____ cm

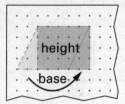

Step 2: Cut out Parallelogram A from Cutout Page 5.1. Cut the shaded triangle from the parallelogram. Move the triangle to the right side of the parallelogram and form a rectangle. Tape the rectangle in the box to the left.

Parallelogram A

base = _____ cm

height = _____ cm

Rectangle

length = _____ cm

width = _____ cm

Area = length × width

     = _____ cm × _____ cm

     = _____ cm²

Step 3: Repeat Step 2 for Parallelogram B.

Parallelogram B

base = _____ cm

height = _____ cm

Rectangle

length = _____ cm

width = _____ cm

Area = _____ cm × _____ cm

= _____ cm²

Reflecting

▸ How do you know that the area of the original parallelogram is the same as the area of the rectangle that was formed?

_____

_____

_____

▸ Use the words "base" and "height" to write a **formula** for calculating the area of a parallelogram.

Area = _____

MATH TERM

formula
a rule represented by symbols, numbers, and/or letters, in the form of an equation; for example, area of a rectangle = length × width

Text page 154

Practising

7. Calculate the area of parallelogram *WXYZ*.

Area = base × height

= _____ m × _____ m

= _____ m²

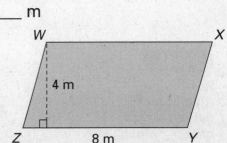

Area of a Triangle

You will need
- Cutout Page 5.2
- scissors
- tape
- a ruler
- a calculator

▶ **GOAL** Develop and use the formula for the area of a triangle.

Use these steps to develop the formula for the area of a triangle.

Step 1: Count the squares inside Δ*ABC* below to estimate the area of the triangle.

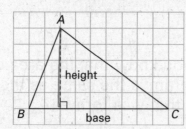

Each square on the grid represents 1 cm^2.

Area = about _____ cm^2

Step 2: Cut out both copies of Δ*ABC* from Cutout Page 5.2. Arrange the two triangles in the box below to form a parallelogram. Tape them to the page.

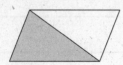

Parallelogram

base = _____ cm

height = _____ cm

Area = _____ cm × _____ cm

= _____ cm^2

Step 3: How much of the parallelogram does the triangle take up? Circle one.

$\frac{1}{2}$ $\frac{1}{4}$ $\frac{1}{3}$

Why does it make sense to say $\frac{1}{2}$ the area of the parallelogram is equal to the area of $\triangle ABC$?

_____

_____

Use the words "base" and "height" to write a formula for calculating the area of a triangle.

Area = _____

Calculate the area of another triangle using the formula you developed above.

Cut out both copies of $\triangle DEF$ on Cutout Page 5.2. Tape the two triangles in the box below to form a parallelogram.

```



```

Parallelogram

base = _____ cm

height = _____ cm

Area = _____ cm $\times$ _____ cm

     = _____ cm^2

$\triangle$DEF

Area = _____ cm^2 $\div$ _____

     = _____ cm^2

TURN ➡

Reflecting

▶ Why can you think of every triangle as half a parallelogram?

_____

_____

Practising

Text page 158

3. Draw a sketch to show how each of these triangles is half a parallelogram. The first one is done for you.

a)

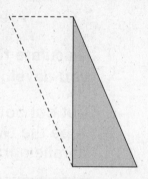

b)

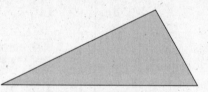

7. Calculate the area of the parallelogram.
   Use this area to calculate the area of the shaded triangle.

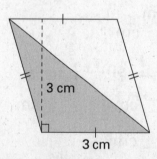

3 cm

3 cm

Parallelogram

Area = _____ × _____

= _____

Triangle

Area = _____ ÷ _____

= _____

8. Measure a base and a height for each triangle.
Then calculate the area.
The height always forms a right angle with the base.

a)

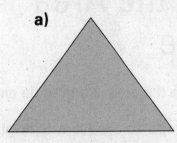

base = _____ cm

height = _____ cm

Area = (_____ × _____) ÷ _____

= _____ ÷ _____

= _____

b)

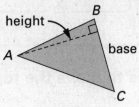

base = _____ cm

height = _____ cm

Area = (_____ × _____) ÷ _____

= _____ ÷ _____

= _____

9. △*ABC* has three different base and height pairs.
One pair is shown in Diagram 1.

Hint

Remember, the height always forms a right angle with the base.

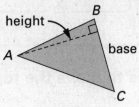

Diagram 1

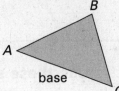

Diagram 2

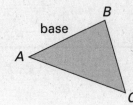

Diagram 3

a) Draw the height that corresponds with the base labelled in Diagram 2 and Diagram 3.

b) When you use a formula to calculate the area of a triangle, does it matter which height-base pair you use? Explain.

_____

_____

_____

Calculating the Area of a Triangle

You will need
• a calculator
• a geoboard
• elastic bands

▶ **GOAL** Explore the area of a triangle on a geoboard.

Use these steps to create a triangle on a geoboard.

Step 1: Select two parallel lines on your geoboard. Mark the parallel lines with elastic bands.

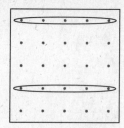

Step 2: Create a triangle that has a base of 3 units and fits exactly between the top and bottom parallel lines. Calculate the area of the triangle.

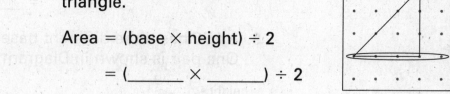

Area = (base × height) ÷ 2

= (_____ × _____) ÷ 2

= _____ units²

Use these steps to move the top vertex of the triangle.

Step 3: Move the top vertex to another position on the top parallel line. Calculate the area of the new triangle.

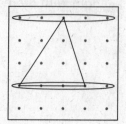

Area = (_____ × _____) ÷ 2

= _____ units²

Step 4: Move the top vertex to another position on the top parallel line. Draw your triangle on the grid. Calculate the area of the new triangle.

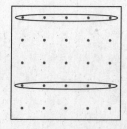

Area = (_____ × _____) ÷ 2

= _____ units²

Use this step to change the base of the triangle.

Step 5: Create two different triangles with a base
of 4 units and calculate their areas.
Draw your triangles on the grids.

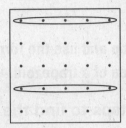 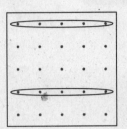

Area = (____ × ____) ÷ 2 Area = (____ × ____) ÷ 2

= ____ units² = ____ units²

Use this step to change the height of the triangle.

Step 6: Select a different bottom parallel line.
Create two different triangles with a base
of 3 units and calculate their areas.
Draw your triangles on the grids.

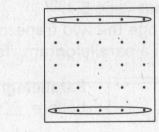 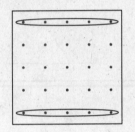

Area = (____ × ____) ÷ 2 Area = (____ × ____) ÷ 2

= ____ units² = ____ units²

Reflecting

▶ Why did the area of the triangle not change when
you moved the vertex in Steps 3 and 4?

_____

_____

▶ What happens to the area of a triangle when you
increase its base or height?

_____

_____

Area of a Trapezoid

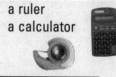

▶ **GOAL** Develop and use the formula for the area of a trapezoid.

Use these steps to find the area of a trapezoid.

Step 1: Count the grid squares and estimate the area of the trapezoid below.

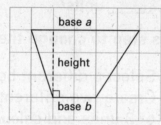

Each square on the grid represents 1 cm².

Area = about _____ cm²

Step 2: Cut out both copies of Trapezoid *ABCD* from Cutout Page 5.4.

Arrange the two trapezoids in the box below to form a parallelogram. Tape them to the page.

Parallelogram

base = _____ cm + _____ cm

= _____ cm

height = _____ cm

Area = _____ cm × _____ cm

= _____ cm²

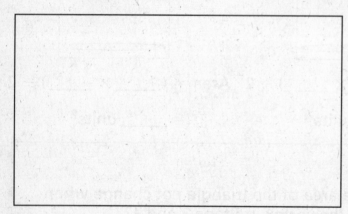

Step 3: How much of the parallelogram does one trapezoid take up? Circle one.

$\frac{1}{2}$ $\frac{1}{4}$ $\frac{1}{3}$

Use the area of the parallelogram to calculate the area of one trapezoid.

Area of Trapezoid *ABCD* = _____ ÷ _____

= _____ cm²

Use these steps to develop the formula for the area of a trapezoid.

Step 1: Cut out both copies of Trapezoid *DEFG* from Cutout Page 5.4.

Arrange the two trapezoids in the box below to form a parallelogram. Tape them to the page.

Parallelogram

base = _____ cm + _____ cm

= _____ cm

height = _____ cm

Area = _____ cm × _____ cm

= _____ cm²

Trapezoid *DEFG*

Area = _____ ÷ _____

= _____ cm²

Step 2: Write out the steps that you used to calculate the area of the trapezoid.

1. Add the _____.

2. Multiply by the _____.

3. Divide by _____.

Step 3: Complete the formula for the area of a trapezoid. Use the letters *a*, *b*, and *h* to represent base *a*, base *b*, and the height *h*.

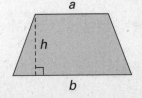

Area = (_____ + _____) × _____ ÷ _____

Reflecting

▶ Why can a trapezoid always be thought of as half of a parallelogram?

_____

_____

_____

Practising

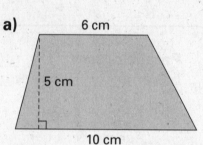

Text page 164 **6.** Calculate the area of each trapezoid using a formula.

a)

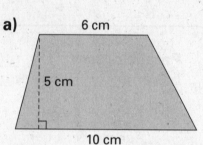

6 cm

5 cm

10 cm

Area = (a + b) × h ÷ 2

b)

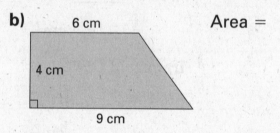

6 cm

4 cm

9 cm

Area =

c)

9 cm

3 cm

2 cm

Area =

9. Measure the trapezoids and calculate their areas.

a)

Area =

b)

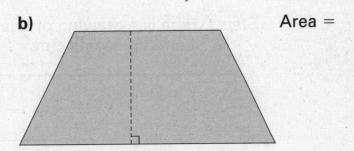

Area =

10. (Circle) the calculation you could use to find the area of the trapezoid below.

a) Area = (3 m + 1 m) × 4 m ÷ 2

b) Area = 1 m + (4 m + 3 m) ÷ 2

c) Area = (4 m + 1 m) × 3 m ÷ 2

d) Area = (1 m + 3 m) × 4 m ÷ 2

e) Area = (4 m + 3 m) × 1 m ÷ 2

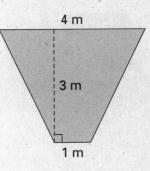

How do you know?

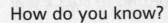

Exploring the Area and Perimeter of a Trapezoid

▶ **GOAL** Explore the relationship between the area and perimeter of a trapezoid.

MATH TERM

isosceles trapezoid
a trapezoid where the non-parallel sides have equal lengths

Brooke is planning a flower garden in the shape of an **isosceles trapezoid**. It will have a perimeter of 24 m.

Use these steps to determine what dimensions will give the garden the greatest area.

Step 1: Here is a drawing of one possible garden. 1 cm represents 1 m.

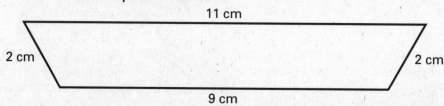

Put your 24 cm string around the outside of the trapezoid to check that the perimeter is 24 cm. You may want to tape down the corners. Measure the height and calculate the area.

height = _____ cm

Area = (11 cm + _____ cm) × _____ cm ÷ _____

= _____ cm²

Hint

Remember,
Area of a trapezoid
= (base *a* + base *b*) × height ÷ 2.

Step 2: Here is a second possible garden. Put your string around the outside of this trapezoid to check that the perimeter is 24 cm.

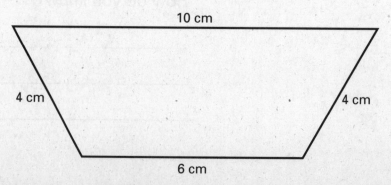

Step 3: Complete the table for the two possible gardens.

Perimeter (cm)	Sketch of possible trapezoid	Side length (cm)	Side length (cm)	Base *a* (cm)	Base *b* (cm)	Height (cm)	Area (cm²)
24	11 cm / 2 cm / 9 cm	2	2	11	9		
24	10 cm / 4 cm / 6 cm	4	4	10			
24							
24							
24							

Hint

Make sure the non-parallel sides of your trapezoid are the same length.

Step 4: Use tape to help arrange your string into three more isosceles trapezoids. Record each trapezoid's measurements in the table. Find their areas.

Step 5: Circle the row of the table for the trapezoid with the greatest area.

Reflecting

▶ What happens to the area of the trapezoid as the sides and height get closer to the same measurement?

_____

_____

END

5.6

Calculating the Area of a Complex Shape

- pattern blocks
- a ruler
- a calculator

▶ **GOAL** Calculate the area of an irregular
2-D shape by dividing it into simpler shapes.

Problem 1

Hint

Use a red trapezoid, a green triangle, and an orange square.

Use these steps to calculate the area of this shape.

Step 1: Cover the shape completely with pattern blocks.
Draw lines on the shape to divide the shape into three simpler polygons.

Step 2: Use a ruler to measure the simpler polygons. Calculate their areas.

Step 3: Complete the table to calculate the total area of the shape using the areas of the simpler polygons.

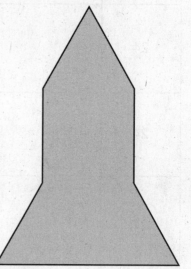

Simpler polygon	Area of simpler polygon
2.0 cm 2.5 cm	$A = b \times h \div 2$ $= 2.5 \times 2 \div 2$ $= 2.5 \ cm^2$
	$A = l \times w$ $=$ $=$
	$A = (a + b) \times h \div 2$ $= (\underline{\hspace{1cm}} + \underline{\hspace{1cm}}) \times \underline{\hspace{1cm}} \div 2$ $=$ $=$ $=$
Total area of shape: $\underline{\hspace{1cm}} + \underline{\hspace{1cm}} + \underline{\hspace{1cm}} = \underline{\hspace{1cm}} \ cm^2$	

112 Lesson 5.6: Calculating the Area of a Complex Shape

Copyright © 2006 Nelson

Use these steps to calculate the area of this shape.

Step 1: Use pattern blocks to divide the shape into simpler polygons. Draw lines on the shape to show the simpler polygons.

Step 2: Use a ruler to measure the simpler polygons.

Step 3: Complete the table to calculate the total area of the shape.

Simpler polygon	Area of simpler polygon
	$A = (a + b) \times h \div 2$ $=$ $=$ $=$ $=$

Total area of shape: _____ + _____ + _____ = _____ cm²

Reflecting

▶ Explain how starting with simpler polygons helps you calculate the area of a complex polygon.

_____

_____

_____

TURN

Practising

Text page 175

6. Calculate the area of the shaded part of each diagram. Show your work in the tables.

a)

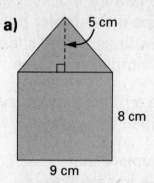

5 cm

8 cm

9 cm

Simpler polygon	Area of simpler polygon
Total area:	

b)

7 cm

2 cm

2 cm

7 cm

This shape has a hole in it. What operation will you need to perform to determine the shaded area?

_____

Simpler polygon	Area of simpler polygon
Total area:	

8. This diagram shows a picnic area.

Each square represents 1 m². The shaded areas are grass. The non-shaded areas need to be paved.

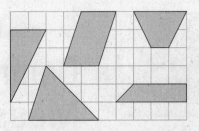

Calculate the total area that needs to be paved. Show your work in the table.

Simpler polygon	Area of simpler polygon
entire picnic area	

Total shaded area: _____

Total area that needs to be paved: _____ − _____ = _____
(entire picnic area) (shaded area)

The paving company charges \$12 to pave 1 m². The total cost to pave the non-shaded area is:

\$12 × _____ = \$_____

END

Communicating about Measurement

You will need
• a hexagon pattern block (optional)

▶ **GOAL** Describe perimeter and area using mathematical language.

Sarah wrote a report to describe how her family used area and perimeter to calculate the cost of outdoor carpet for their deck.
The carpeted area is 10 m long and 2 m wide.
The carpet costs $1.25/m².
The moulding around the carpet costs 50¢/m.

10 m
2 m

Rosa asked questions to help Sarah improve her report.
Read Sarah's report and answer Rosa's questions.

Rosa's Questions

1. How do you measure the perimeter?

_____

_____

2. Why do you need to calculate the area?

_____

_____

3. Why do you use the length and the width to calculate the area?

_____

_____

4. How do you know which price to use for each measurement?

_____

_____

Sarah's Report

First I measured the perimeter. I got 24 m.

Next I calculated the area.

I used length × width. The area equals 20.

To calculate the amount of moulding, I multiplied 24 by $0.50.

To calculate the amount of carpet, I multiplied 20 by $1.25. The total cost is $37.

Communication Checklist

❑ Did you include all the important details?

❑ Did you use correct units, symbols, and vocabulary?

❑ Did you completely solve the problem?

Reflecting

▶ Which part of the Communication Checklist did Sarah cover well? Explain your answer.

_____

_____

_____

_____

Practising

Text page 178

Hint

If you don't have a hexagon pattern block, you can measure the hexagon on this page.

Hint

What simpler shapes make up the hexagon?

5. Calculate the area of the top face of a yellow hexagon pattern block.
Write a report to explain what you did.
Use the Communication Checklist to help you.

Step 1: Explain how you plan to solve the problem.

_____

_____

_____

_____

Step 2: Solve the problem. Show your work.

The area of the hexagon is _____.

Copyright © 2006 Nelson

Lesson 5.7: Communicating about Measurement **117**

END

Comparing Positive and Negative Numbers

▶ **GOAL** Compare and order positive and negative numbers.

Think of a number line as a thermometer.
They both have positive and negative **integers** marked on them.

MATH TERM

integers
all positive and
negative whole
numbers:
... −3, −2, −1, 0, +1,
+2, +3, ...

A. Circle the correct word. Look at the thermometer and number line below for help.

Positive integers are at the **top / bottom** of a thermometer.
Positive integers are to the **left / right** of zero on a number line.

Negative integers are at the **top / bottom** of a thermometer.
Negative integers are to the **left / right** of zero on a number line.

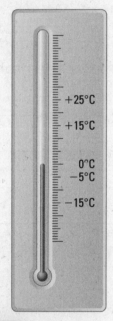

B. Transfer the temperatures from the thermometer to the number line below.

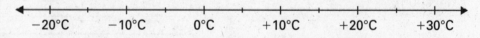

−20°C   −10°C   0°C   +10°C   +20°C   +30°C

C. Circle the correct statement. The first one is done for you. Use the thermometer and number line in Part B for help. When numbers are greater, they are farther right on a number line. For example, −10 is greater than −20.

Hint

Remember,
> means "greater than" and
< means "less than."

−20°C > +15°C	OR	⟨−20°C < +15°C⟩
0°C > −5°C	OR	0°C < −5°C
−15°C > −10°C	OR	−15°C < −10°C
−10°C > +5°C	OR	−10°C < +5°C

Reflecting

▶ How can a thermometer help you list integers in order from least to greatest?

_____

_____

_____

▶ How can using a number line help you list integers in order from least to greatest?

_____

_____

_____

Practising

Text page 187

4. Circle the integer that is greater.

a) +3   OR   +2

b) −3   OR   +2

c) +3   OR   −2

d) −3   OR   −2

5. Mark the following integers on the number line.

a) +8, −7, −2, +1, +3, −5

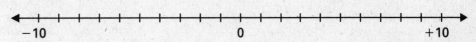

−10   0   +10

b) Write the integers in order from least to greatest.

_____

TURN

8. Mark the integers on the number line. Then write the integers in order from least to greatest.

a) −3, +7, −8, +2

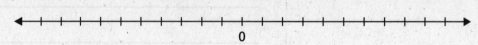

Least to greatest: _____

b) +35, +15, −20, −10, +5, −40

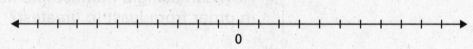

Least to greatest: _____

9. a) Mark each integer below on the number line.

+1, −4, +3, −3, −1, −9, +8

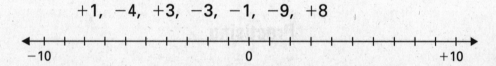

b) Write **<** or **>** to make each statement true.

+1 ☐ −4

+3 ☐ −3

−1 ☐ 0

+8 ☐ −9

10. Mark each integer described below on the number line.

a) three greater than zero

b) four less than zero

c) two greater than positive four

d) five less than negative two

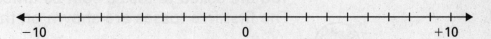

Text page 189 **16.** A basketball coach rated each player's skill.

Name	Score
Jan	−2
Raj	+3
Toni	+3
Monica	−1
Arvin	−4
Ming	+4
Barbara	+3
Riki	−2

a) The five players with the highest scores will start the next game. Name the starting players.

_____

_____

b) The player with the highest score will be captain. Name the captain.

_____

END

6.2

An Integer Experiment

You will need
- a coin

▶ **GOAL** Add positive and negative numbers.

You are asked to toss a coin 20 times as an experiment. The first five tosses are shown on the table below.

Heads (H) means you gain 1 point (+1).

Tails (T) means you lose 1 point (−1).

A. Toss a coin 15 MORE times and complete the table for 20 tosses.

Toss number	Result H or T	Point value (+1) or (−1)	Total score
1	T	−1	−1
2	T	−1	−2
3	H	+1	−1
4	H	+1	0
5	H	+1	+1
6			
7			
8			
9			
10			
11			
12			
13			
14			
15			
16			
17			
18			
19			
20			

B. Plot your results for each toss on the scatter plot. Use ○ for Heads and ◉ for Tails.
Draw a dashed line to show how your total score changes with each toss. The first five tosses are done for you.

Hint

Remember, every time the coin lands Heads, the dashed line goes up one.

Every time the coin lands Tails, the dashed line goes down one.

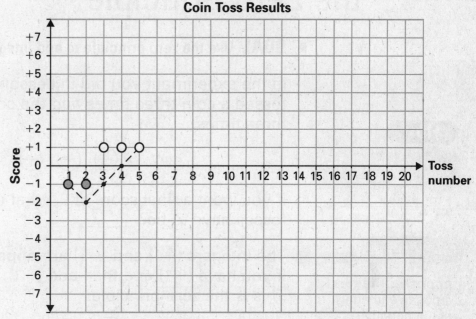

Coin Toss Results

C. What was the total score after five tosses? _____
How does the graph show this?

_____

_____

D. What is your final score? _____
How can you tell just by looking at your graph whether your score is positive or negative?

_____

_____

Reflecting

► What happens to the total score when the value of the next toss is (−1)?

_____

► What happens to the total score when the value of the next toss is (+1)?

_____

Adding Integers Using the Zero Principle

You will need
- red and blue counters

▶ **GOAL** Use the zero principle to add integers.

In the experiment you did in Lesson 6.2, suppose you tossed a coin three times and the coin landed Heads three times.

$\oplus$ $\oplus$ $\oplus$

If you want a final score of 0, what do your next three tosses have to be? _____

Hint

Heads = (+1) or $\oplus$

Tails = (−1) or $\ominus$

The integers (+1) and (−1) are **opposite integers**.
If you have (+1) and then add (−1), you have zero.
This is the zero principle.

MATH TERM

opposite integers
a positive integer and a negative integer that are the same distance from zero; for example, +6 and −6 are opposite integers

−6 −4 −2 0 +2 +4 +6

Model (+3) + (−3) using counters.
Pair each red counter (+1) with a blue counter (−1).
Each pair of red and blue counters equals 0.

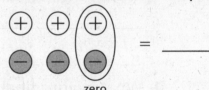

= _____

Follow these instructions to see how different models of the same number can be made using the zero principle.

A. Write the integer modelled by each set of counters in the table.

Model	Integer
$\oplus$ $\oplus$ $\oplus$ $\oplus$	
$\oplus$ $\oplus$ $\oplus$ $\oplus$ $\oplus$ $\ominus$	
$\oplus$ $\oplus$ $\oplus$ $\oplus$ $\oplus$ $\ominus$ $\oplus$ $\ominus$	

B. Why do all the models in the table on page 124 represent the same integer?

_____

_____

_____

Here is a model of (+5) + (−3) using counters and the zero principle.

= ⊕⊕⊕⊕⊕ + ⊖⊖⊖

= (⊕⊖) (⊕⊖) (⊕⊖) ⊕⊕

= ⊕⊕

So, (+5) + (−3) = (+2).

Use counters to model (−6) + (+3). Draw your model below. Then write the sum using numbers.

=

=

=

(_____) + (_____) = (_____)

Hint

(⊖⊕) = zero

Reflecting

▶ The sum of any two opposite integers is always zero. True or false? _____
Verify your answer using counters.
Give an example to support your answer.
(_____) + (_____) = _____

▶ Draw three more models of (−2) using counters.

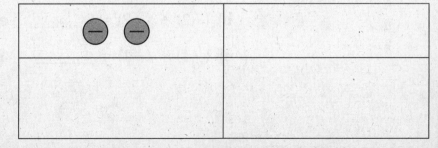

Hint

Use the zero principle.

Practising

Text page 194 **3.** Represent each expression using counters.
Draw your models in the table.
Then write the sum as a number.
The first one is done for you.

Expression	Counter model	Sum (number)
a) (−3) + (+2)	= ⊖⊖⊖ + ⊕⊕ = (⊖⊕) (⊖⊕) ⊖ = ⊖	(−1)
b) (−4) + (+6)		
c) (+5) + (−6)		
d) (+3) + (−3)		

5. Complete. Use counters to help you add.

a) (−3) + (−2) = _____   **d)** (−7) + (+6) = _____

b) (+2) + (−2) = _____   **e)** (−5) + (−2) = _____

c) (−4) + (+1) = _____   **f)** (−5) + (+2) = _____

6. Explain why $(-25) + (+25) = 0$.

_____

_____

9. Complete each question using numbers.
The first one is done for you.

a) ⊖ ⊖ ⊖   ⊕ ⊕ ⊕   ⊕ ⊕ ⊕ ⊕ ⊕

  (-3) + $(+3)$ + $(+5)$ = $(+5)$

b) ⊖   ⊕ ⊕   ⊖

  _____ + _____ + _____ = _____

c) ⊕ ⊕   ⊖ ⊖ ⊖   ⊕ ⊕ ⊕ ⊕ ⊕ ⊕

  _____ + _____ + _____ = _____

Text page 195 **13. a)** Use counters to create three different mode⸝ of
(-4). Draw your counter models below. The ⸝st
one is done for you.

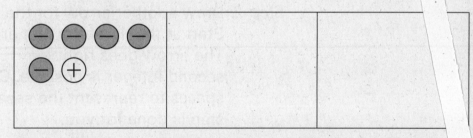

b) Use your models to fill in each blank with an
integer to make the equation true.

  $(\underline{}) + (\underline{}) + (\underline{}) = (-4)$

  $(\underline{}) + (\underline{}) + (\underline{}) = (-4)$

  $(\underline{}) + (\underline{}) + (\underline{}) = (-4)$

END

Adding Integers That Are Far from Zero

▶ **GOAL** Add integers using a number line.

Problem 1

You are asked to calculate $(-10) + (+30)$, but you don't have enough counters.

Use these steps to calculate $(-10) + (+30)$ using a number line.

Hint

positive integers

negative integers

Step 1: Draw an arrow for the first integer in the addition sentence. Start at 0.
This step is done for you.

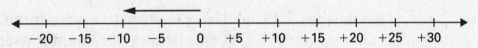

Step 2: Draw another arrow for the second integer. Start at the tip of the first arrow.
The arrow goes right (⟶) because the second integer is positive. Count the number of spaces to represent the second number. This step is done for you.

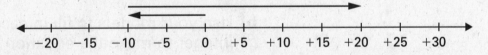

Step 3: The second arrow ends on the number line at the sum of the integers.

So, $(-10) + (+30) = ($_____$)$.

Problem 2

Use these steps to identify an integer addition equation on a number line.

Step 1: The bottom arrow starts at 0 so it is the first number in the equation.
Write the number that the bottom arrow represents.

_____

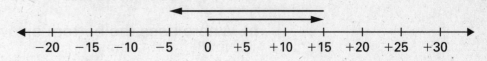

−20 −15 −10 −5   0   +5 +10 +15 +20 +25 +30

> **Hint**
>
> Remember, arrows that go left (←———) represent negative integers.

Step 2: Count the number of spaces that the top arrow crosses. Write the number that the top arrow represents.

_____

Step 3: Write the number where the top arrow ends.

_____

Step 4: Write the equation.

(_____) + (_____) = (_____)

Reflecting

▶ How can you predict whether the sum of two integers will be positive or negative, without adding the integers?

_____

_____

_____

_____

▶ Show how you would model the **zero principle** on a number line.

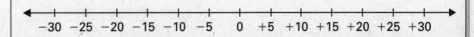

−30 −25 −20 −15 −10 −5   0   +5 +10 +15 +20 +25 +30

> **MATH TERM**
>
> **zero principle**
> when two opposite integers are added, they give a sum of zero

TURN ➤

Practising

Text page 198

5. Use the number line below to model the sum
  (−25) + (+35).

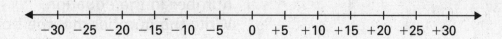

a) Where does the bottom arrow start? _____

b) Where does the bottom arrow end? _____

c) Where does the top arrow start? _____

d) Where does the top arrow end? _____

e) What is the sum?

  (−25) + (+35) = (_____)

Text page 199

9. Add. Draw a number line model for each sum.

a) (+5) + (−10) = (_____)

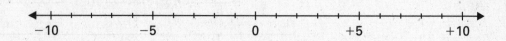

b) (+10) + (−5) = (_____)

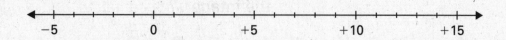

c) (−5) + (−10) = (_____)

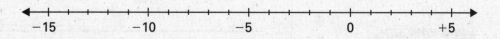

12. a) The sum of two negative integers is always negative. Model an example on the number line to show this is true.

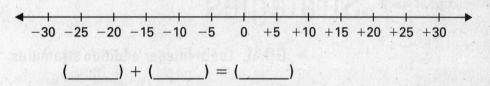

(_____) + (_____) = (_____)

b) If the sum of two integers is zero, the two integers must be opposites. Model an example to show this is true.

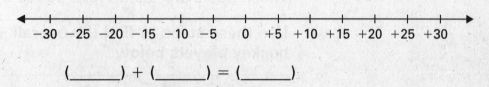

(_____) + (_____) = (_____)

c) The sum of a positive integer and a negative integer is negative. Model an example to show this is *not* always true.

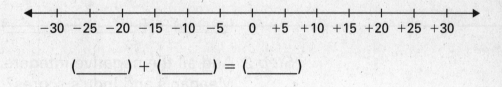

(_____) + (_____) = (_____)

15. Complete the table.

Starting temperature (°C)	Temperature change (°C)	Final temperature (°C)	Number line model
−5	+1		
−10	−4		
+7		−2	

END

Integer Addition Strategies

▶ **GOAL** Learn integer addition strategies.

A hockey player's +/− score is determined by the goals scored while on the ice.

• A goal scored for the team counts as (+1).

• A goal scored against the team counts as (−1).

These goals are added to produce a +/− score.

Use these steps to find the overall +/− score for the hockey players below.

Heidi	Rana	Meagan	Sonya	Indu
+16	+14	−9	+8	−1

Step 1: Add all the positive integers. What is the sum of Heidi's, Rana's, and Sonya's scores?

(_____) + (_____) + (_____) = (_____)

Step 2: Add all the negative integers. What is the sum of Meagan's and Indu's scores?

(_____) + (_____) = (_____)

Step 3: Add the sums of the positive and negative integers. What is the overall +/− score for the five players?

(_____) + (_____) = (_____)

Hint

Add the sums in Steps 1 and 2. Draw a number line on scrap paper to help you add.

Step 4: Using your calculator, add:

(+16) + (+14) + (−9) + (+8) + (−1) = (_____)

To calculate the sum, press these keys:

16 + 14 + 9 +/− + 8 + 1 +/− =

Hint

Press the +/− key after a negative integer to change the sign.

Reflecting

▶ Why are the answers in Steps 3 and 4 the same?

_____

_____

_____

Practising

Text page 202

4. Group all the positive integers and negative integers.
   Calculate the sum of each group.
   Then calculate the total sum.

$$(-13) + (+8) + (-12) + (+10) + (+9)$$

Hint

Draw a number line on scrap paper to help you add.

=

=

=

5. Explain why these expressions all have the same sum.

   a) $(-5) + (-2) + (-3) + (+5)$

   b) $(-5) + (+5) + (-2) + (-3)$

   c) $(-3) + (-2) + (+5) + (-5)$

   d) $(+5) + (-2) + (-3) + (-5)$

   _____

   _____

   _____

Text page 203

7. Calculate the sum. Check the answer with a calculator.

$$(-12) + (+2) + (-5)$$

Check answer

(_____)

=

=

=

Using Counters to Subtract Integers

▶ **GOAL** Subtract integers using counters.

Here is a drawing of a counter model of (+5) − (+3).

Hint

Remember, subtract means take away.

The counters that are subtracted, or taken away, are crossed out.

Complete the model of (−4) − (−3).
Cross out the counters that you are taking away.

Why can you not model (−4) − (+3) using the model above?

_____

Use these steps to model differences by adding zeros.

Step 1: Represent the integer (−4) using counters.
Draw your model below.

Step 2: Use the zero principle to make a different model
of (−4) that includes three red (positive)
counters. Draw your model below.

Hint

To make the model still equal 3, add zeros.

  = zero

Now you have positive counters that you can take away.

Step 3: Use your model in Step 2 to answer:

(−4) − (+3) = (_____)

Step 4: Add as many zeros as you need depending on the subtraction question.

To model $(-4) - (+5)$, draw a model of (-4) that includes five zeros so you have five positive counters that you can take away.

Step 5: Use your model in Step 4 to answer:

$$(-4) - (+5) = (\underline{\hspace{3em}})$$

Reflecting

▶ Why do you *not* need to add zeros to model $(-4) - (-3)$?

_____

_____

▶ Write another integer subtraction question that does *not* require adding zeros.

$$(\underline{\hspace{2em}}) - (\underline{\hspace{2em}})$$

▶ Why do you need to add zeros to model $(-4) - (+3)$?

_____

_____

▶ Write another integer subtraction question that requires adding zeros.

$$(\underline{\hspace{2em}}) - (\underline{\hspace{2em}})$$

▶ How many zeros must be added to your question above?

_____

Practising

Text page 210

6. Use counters to model each question in the table.
Identify whether or not the model requires zeros.
Then complete the table.
The first one is done for you.

	Subtraction	Zeros (Yes/No)	Number of zeros
a)	$(-4) - (+2) = (-6)$	Yes	2
b)	$(+3) - (+2) = (\underline{})$		
c)	$(+3) - (-2) = (\underline{})$		
d)	$(-3) - (-2) = (\underline{})$		
e)	$(-2) - (-3) = (\underline{})$		

7. a) Use counters to calculate $(-4) - (-1) = (\underline{})$.
Draw your model here.

b) Why do you not need to add zeros to complete this subtraction?

_____

_____

Hint

How many more negative counters do you need to subtract four negative counters?

8. a) Use counters to calculate $(-1) - (-4) = (\underline{})$.
Draw your model here.

b) Why did you need to add zeros to complete this subtraction?

_____

_____

Hint

See Lesson 6.3 on page 124 to recall how to model addition of integers using counters.

9. Draw a counter model for $(-3) - (+4)$ and another counter model for $(-3) + (-4)$. The models should show that $(-3) - (+4) = (-3) + (-4)$.

$(-3) - (+4) = ($_____$)$	$(-3) + (-4) = ($_____$)$

Text page 211 **19.** Is each statement true or false?

a) Model each difference using counters.

$(-4) - (-3) = ($_____$)$ $\qquad$ $(-3) - (-4) = ($_____$)$

"The difference between two negative numbers is *always* negative."
Circle the correct answer.

$\qquad\qquad$ True $\qquad$ OR $\qquad$ False

b) Model each difference using counters.

$(+6) - (+2) = ($_____$)$ $\qquad$ $(+2) - (+6) = ($_____$)$

"The difference between two positive numbers is *always* positive."
Circle the correct answer.

$\qquad\qquad$ True $\qquad$ OR $\qquad$ False

c) Model each difference using counters.

$(+4) - (-3) = ($_____$)$ $\qquad$ $(+1) - (-4) = ($_____$)$

"When you subtract a negative number from a positive number, the difference is *always* positive."
Circle the correct answer.

$\qquad\qquad$ True $\qquad$ OR $\qquad$ False

Using Number Lines to Subtract Integers

▶ **GOAL** Subtract integers using a number line.

On a number line, subtraction is represented by the (ending position) − (starting position) of the arrow.

The arrow on the number line below starts at (−3) and ends at (+6).

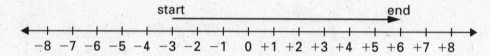

The number line models (+6) − (−3).

To find the difference, count the spaces between the ending and starting position.

(+6) − (−3) = (_____)

Hint

The direction of the arrow tells you if the difference is positive or negative.

positive integers

negative integers

Fill in the blanks to name the integer subtraction the number line represents.

The arrow on the number line below starts at ☐ and ends at ☐ .

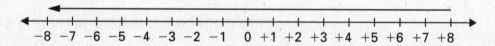

The number line models the subtraction ☐ − ☐ .

The difference is ☐ .

Reflecting

Use this number line to help you answer the questions below.

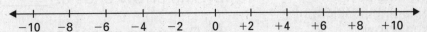

▶ How is an arrow going from (−4) to (+6) the same as an arrow going from (+6) to (−4)?

_____

▶ How is an arrow going from (−4) to (+6) different from an arrow going from (+6) to (−4)?

_____

_____

_____

Practising

Text page 214

4. Write the subtraction question that each model represents. Calculate each difference.

a)

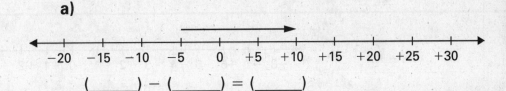

(_____) − (_____) = (_____)

b)

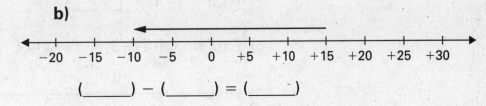

(_____) − (_____) = (_____)

c)

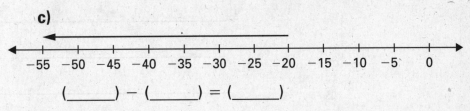

(_____) − (_____) = (_____)

Hint

Remember, on a number line, subtraction is represented by the (ending position) − (starting position) of the arrow.

6. An arrow is used on a number line model to represent $(-35) - (+40)$.

a) The arrow starts at _____.

b) The arrow ends at _____.

c) Draw the arrow for this model.

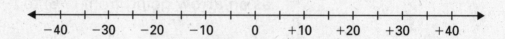

d) The difference is $(-35) - (+40) = ($_____$)$.

Text page 215 **9. a)** Use the number line below to model the difference $(-3) - (+4)$.

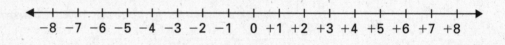

$(-3) - (+4) = ($_____$)$

b) Use the number line below to model the sum $(-3) + (-4)$.

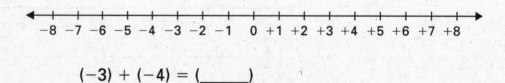

$(-3) + (-4) = ($_____$)$

c) Explain why $(-3) - (+4) = (-3) + (-4)$.

_____

_____

_____

_____

11. Model each difference on the number line.
Write the difference in the equation.
The first one is done for you.

a) $(-20) - (-40) = (+20)$

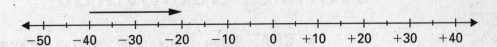

b) $(+30) - (+70) = ($_____$)$

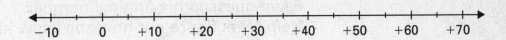

c) $(-23) - (-21) = ($_____$)$

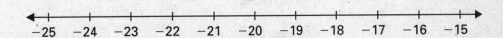

d) $(+10) - (-10) = ($_____$)$

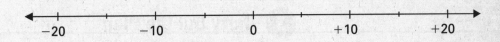

Solve Problems by Working Backwards

▶ **GOAL** Solve problems by working backwards.

Fawn asked Mark to think of a number and then follow
the steps in the first table below.
Mark told Fawn his result.
Fawn guessed his original number.
Help Mark figure out how she knew his original number.

❶ Understand the Problem

Mark wants to know how Fawn knew his original
number.

❷ Make a Plan

Mark decides he needs to work backwards through the
steps to find the original number.

❸ Carry Out the Plan

Mark follows Fawn's steps in the table.

Step 1: Think of a number.	Step 2: Add (−3) to the number.	Step 3: Subtract the result from 6.	Step 4: Write the opposite of the result.	Step 5: Add (+7) to the result.
+5 →	+2 →	+4 →	−4 →	+3

Then he rewrites the steps to work backwards from the
result.

Step 5: Subtract (−3) from the result.	Step 4: Subtract the result from 6.	Step 3: Write the opposite of the result.	Step 2: Subtract (+7) from the result.	Step 1: Start with the number you ended with.
←	←	←	←	+3

Work through the steps in the second table.
Did you get the original number? _____

Reflecting

▶ How did working backwards help solve the problem?

_____

_____

_____

Practising

Text page 218

3. Try Fawn's steps with three other numbers.

	Step 1: Think of a number.	Step 2: Add (−3) to the number.	Step 3: Subtract the result from 6.	Step 4: Write the opposite of the result.	Step 5: Add (+7) to the result.
a)					
b)					
c)					

Look at the first and last columns.
Explain a quick way to find the original number.

Hint

Think about the relationship between the original number and the result.

_____

_____

Hint

On Thursday, Heidi had two times as many grapes as she did on Friday.

10. On Monday, Heidi bought some grapes. Each day, she ate half of them. On Friday, only 8 grapes were left. How many grapes did Heidi buy?

Work backwards from Friday to solve the problem.

Monday	Tuesday	Wednesday	Thursday	Friday
←	←	←	←	8

END

▶ **GOAL** Locate positions on a Cartesian grid with integer coordinates.

MATH TERM

Cartesian coordinate system
a grid that allows you to describe a location using an ordered pair of coordinates, such as (−3, 4)

The map below uses a **Cartesian coordinate system.**
Bloomsberg is located at (−3, 4) on the map.
From the origin, go left to −3 on the horizontal axis, then up to 4 on the vertical axis.

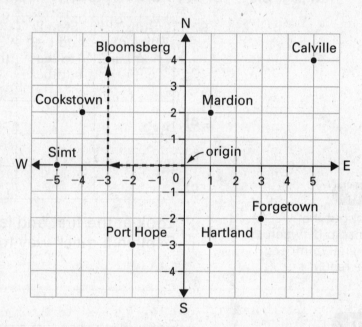

Hint

Find the first number in the brackets on the horizontal axis, and the second number on the vertical axis.

A. Use the map to name the location represented by each set of coordinates.

(5, 4) _____   (−5, 0) _____

(−2, −3) _____   (1, −3) _____

B. Name the coordinates for each of the locations on the map above.

Forgetown   (3, _____)

Mardion   (_____, 2)

Cookstown   (_____, _____)

Hint

The axes on a Cartesian coordinate system are like a compass.

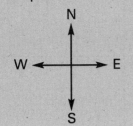

C. Plot the following locations on the map on page 144.

Margotbend   (3, 1)
Janecentre   (2, 4)

Which location is farther north? _____

How can you tell by just looking at the numbers?

_____

_____

_____

_____

Hint

The first number in the pair of coordinates tells you how far east or west the point is.

Reflecting

▶ You are given the coordinates of two points on a grid. How can you tell which point is farther east?

_____

_____

_____

Practising

Text page 230

4. Name the coordinates for each point on the grid.

A(_____, _____)

B(_____, _____)

C(_____, _____)

D(_____, _____)

E(_____, _____)

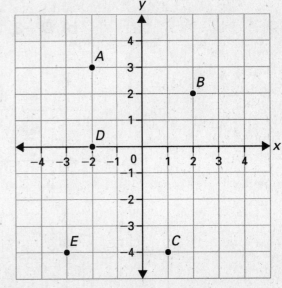

5. Plot the following points on the Cartesian coordinate system.

$A(2, 3)$

$B(-1, 4)$

$C(-3, -5)$

$D(0, 0)$

$E(4, -5)$

$F(0, -4)$

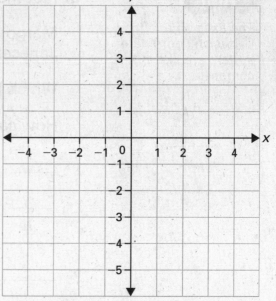

Hint

Remember, find the first coordinate on the horizontal axis. Then go up to the second coordinate on the vertical axis.

6. Plot the set of points on the grid. Connect the points in order. Join the last point to the first point.

a) $A(0, 5)$, $B(4, 5)$, $C(4, 0)$

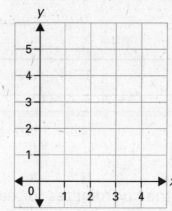

Name the polygon.

_____

b) $D(-3, 1)$, $E(-3, -3)$, $F(-1, -4)$, $G(-1, 3)$

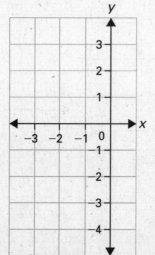

Name the polygon.

_____

Hint

Remember,
< means less than
and > means greater
than.

8. Fill in each blank with a number to make the statement true. Look at the grid at the bottom of the page for help.

a) (5, −3) is to the right of (_____, −3) because 5 > _____.

b) (4, 0) is below (4, _____) because 0 < _____.

11. Name a point that fits each description. Look at the grid at the bottom of the page for help.

a) below (4, −2)

   (_____, _____)

b) to the left of (4, −2)

   (_____, _____)

Text page 231

18. Each unit on this map represents 5 km.

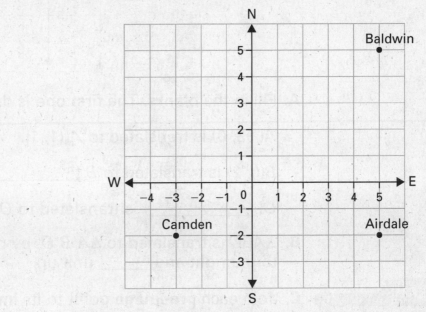

a) Find the distance between Baldwin and Airdale.

   _____ × 5 km = _____ km

b) Find the distance between Camden and Airdale.

   _____ × 5 km = _____ km

Translations

You will need
- a ruler
- coloured pencils

▶ **GOAL** Recognize the image of a 2-D shape after a translation.

MATH TERM

translate
to slide an object along straight lines, left or right, up or down

△*ABO* is **translated** to triangle △*A′B′O′*.

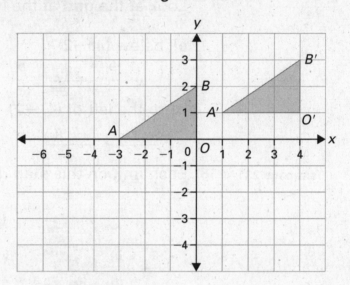

A. Fill in the blanks. The first one is done for you.

$A(-3, 0)$ is translated to $A'(1, 1)$.

$B(0, 2)$ is translated to $B'(\rule{1cm}{0.15mm}, \rule{1cm}{0.15mm})$.

$O(\rule{1cm}{0.15mm}, \rule{1cm}{0.15mm})$ is translated to $O'(\rule{1cm}{0.15mm}, \rule{1cm}{0.15mm})$.

B. △*ABO* is translated to △*A′B′O′* by moving _____ units to the right and _____ unit up.

MATH TERMS

pre-image
the original shape or point

image
the new shape or point created by a transformation; for example, the image of point *A* is *A′*

C. Join each **pre-image** point to its **image** point with a straight line in a different colour. What do the lines have in common?

_____

D. Translate △*ABO* 3 units to the left and 4 units down on the grid above. Name the coordinates of the vertices of the new triangle.

$A''(\rule{1cm}{0.15mm}, \rule{1cm}{0.15mm})$, $B''(\rule{1cm}{0.15mm}, \rule{1cm}{0.15mm})$, $O''(\rule{1cm}{0.15mm}, \rule{1cm}{0.15mm})$

Reflecting

▶ Compare the pre-image and the translated images on page 148. What properties of a pre-image are the same and what properties are different after a translation? Include **orientation**, shape, size, and location in your answer.

_____

_____

_____

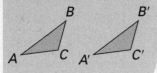
Practising

Text page 234

7. a) Describe the translation that moved *EFGHI* to *E'F'G'H'I'*.

Move *EFGHI* _____ unit(s) down.

Name the coordinates of the vertices.

$E(-1, 3) \rightarrow E'(-1,$ _____ $)$

$F($ _____ , _____ $) \rightarrow F'($ _____ , _____ $)$

$G($ _____ , _____ $) \rightarrow G'($ _____ , _____ $)$

$H($ _____ , _____ $) \rightarrow H'($ _____ , _____ $)$

$I($ _____ , _____ $) \rightarrow I'($ _____ , _____ $)$

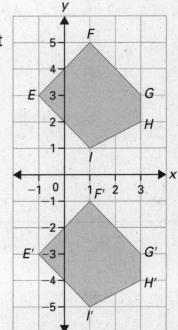

b) Describe the translation that moved *ABCD* to *A'B'C'D'*.

Move *ABCD* _____ unit(s) to the _____ and _____ unit(s) _____ .

Name the coordinates of the vertices.

$A(2, 2) \rightarrow A'($ _____ , _____ $)$

$B($ _____ , _____ $) \rightarrow B'($ _____ , _____ $)$

$C($ _____ , _____ $) \rightarrow C'($ _____ , _____ $)$

$D($ _____ , _____ $) \rightarrow D'($ _____ , _____ $)$

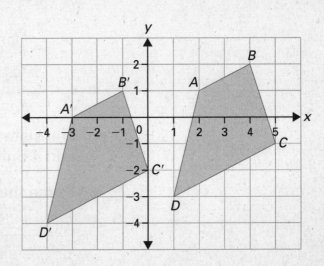

8. Translate △*ABC* 3 units to the left and 2 units up.
Name the coordinates of the vertices.

A(_____, _____) → *A*′(_____, _____)

B(_____, _____) → *B*′(_____, _____)

C(_____, _____) → *C*′(_____, _____)

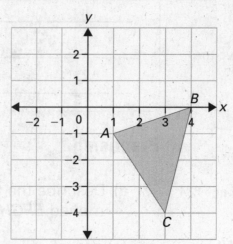

Text page 235 **9. a)** Plot parallelogram *ABCD* on the grid.

A(−5, 3), *B*(0, 2), *C*(1, −1), *D*(−4, 0)

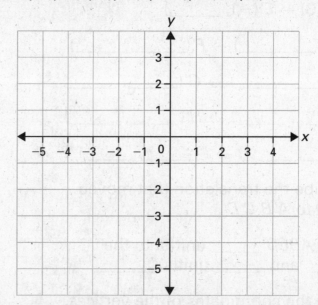

b) Translate parallelogram *ABCD* 3 units to the right
and 4 units down. Label the image *A*′*B*′*C*′*D*′.

c) Name the coordinates of the image *A*′*B*′*C*′*D*′.

A′(_____, _____), *B*′(_____, _____),

C′(_____, _____), *D*′(_____, _____)

10. a) Draw any triangle on the grid.
Label the triangle *DEF.*
Name the coordinates of each vertex.

D(_____, _____)

E(_____, _____)

F(_____, _____)

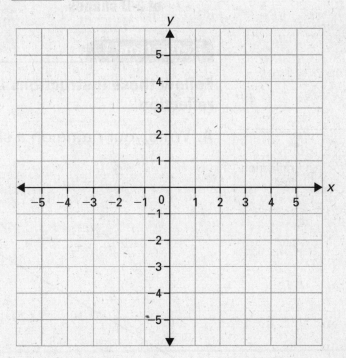

b) Translate Δ*DEF* 5 units to the left and 4 units down. Label the image Δ*D′E′F′*.

c) Name the coordinates of the image Δ*D′E′F′*.

D′(_____, _____)

E′(_____, _____)

F′(_____, _____)

Reflections

You will need
- a ruler
- a transparent mirror
- a protractor
- coloured pencils

▶ **GOAL** Explore the properties of reflections of 2-D shapes.

Reflection 1

Follow these instructions to explore the properties of a reflection.

A. Write your name on a slant above the reflection line.

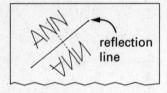

reflection line

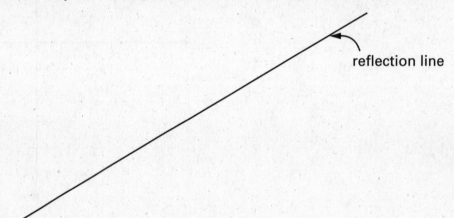

reflection line

B. Place a transparent mirror on the reflection line and use it to draw the image of your name.

C. With a different colour, connect any three points on your name with their image points.

D. Use a protractor to measure the angle between the reflection line and the coloured lines that you drew in Part C. Write the angle.

_____

E. Compare the pre-image to the reflected image. What properties of a pre-image are the same and/or different after a reflection?

Hint

Include shape, size, and distance to the reflection line in your answer.

_____

_____

Follow these instructions to explore the properties of a reflection on a Cartesian grid.

A. $\triangle ABC$ is reflected in the y-axis to create $\triangle A'B'C'$. Place your transparent mirror on the y-axis to check.

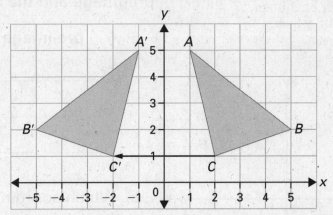

B. Name the coordinates of each vertex and its image.

pre-image   **image**

$A(\_\_\_\_, \_\_\_\_) \rightarrow A'(\_\_\_\_, \_\_\_\_)$

$B(\_\_\_\_, \_\_\_\_) \rightarrow B'(\_\_\_\_, \_\_\_\_)$

$C(\_\_\_\_, \_\_\_\_) \rightarrow C'(\_\_\_\_, \_\_\_\_)$

C. What do you notice about the x-coordinate of each point on the pre-image and the image?

_____

_____

What do you notice about the y-coordinate of each point?

_____

Hint

Remember, orientation means the direction that a shape is facing.

Reflecting

▶ How does the orientation of a pre-image compare to the orientation of an image after a reflection?

_____

_____

Practising

Text page 238

Hint

$F(-2, 3)$ is 3 units above the y-axis, so F' will be 3 units below the y-axis.

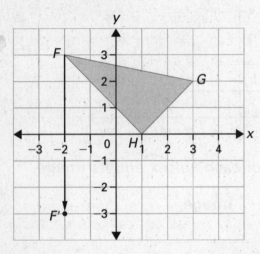

6. a) Reflect △FGH in the x-axis. This means the x-axis will be the reflection line.

b) Name the coordinates for the vertices of the pre-image and the image.

pre-image   **image**

$F(\_\_\_\_, \_\_\_\_) \rightarrow F'(\_\_\_\_, \_\_\_\_)$

$G(\_\_\_\_, \_\_\_\_) \rightarrow G'(\_\_\_\_, \_\_\_\_)$

$H(\_\_\_\_, \_\_\_\_) \rightarrow H'(\_\_\_\_, \_\_\_\_)$

c) Which coordinates stayed the same? Circle one.

x-coordinates   y-coordinates

8. a) Reflect pentagon $DEFGH$ in the y-axis.

b) Name the coordinates for the vertices of the pre-image and the image.

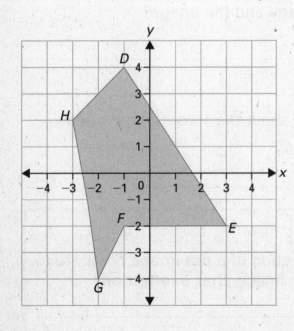

pre-image   **image**

$D(\_\_\_\_, \_\_\_\_) \rightarrow D'(\_\_\_\_, \_\_\_\_)$

$E(\_\_\_\_, \_\_\_\_) \rightarrow E'(\_\_\_\_, \_\_\_\_)$

$F(\_\_\_\_, \_\_\_\_) \rightarrow F'(\_\_\_\_, \_\_\_\_)$

$G(\_\_\_\_, \_\_\_\_) \rightarrow G'(\_\_\_\_, \_\_\_\_)$

$H(\_\_\_\_, \_\_\_\_) \rightarrow H'(\_\_\_\_, \_\_\_\_)$

12. a) Reflect △*MAP* in the *x*-axis.

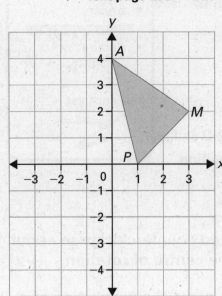

b) Name the coordinates of the vertices of the pre-image and the image.

          pre-image   **image**

$M(\underline{\hspace{1cm}}, \underline{\hspace{1cm}}) \rightarrow M'(\underline{\hspace{1cm}}, \underline{\hspace{1cm}})$

$A(\underline{\hspace{1cm}}, \underline{\hspace{1cm}}) \rightarrow A'(\underline{\hspace{1cm}}, \underline{\hspace{1cm}})$

$P(\underline{\hspace{1cm}}, \underline{\hspace{1cm}}) \rightarrow P'(\underline{\hspace{1cm}}, \underline{\hspace{1cm}})$

c) Reflect △*M′A′P′* in the *y*-axis.

d) Name the coordinates of the vertices of △*M″A″P″*.

$M''(\underline{\hspace{1cm}}, \underline{\hspace{1cm}}) \quad A''(\underline{\hspace{1cm}}, \underline{\hspace{1cm}})$

$P''(\underline{\hspace{1cm}}, \underline{\hspace{1cm}})$

15. Quadrilateral *A′B′C′D′* is a reflection of *ABCD*.

a) Draw the reflection line.

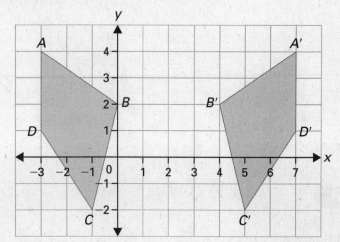

b) How do you know this line is the reflection line? Give two reasons.

_____

_____

_____

_____

_____

END

7.4 Rotations

Text page 240

▶ **GOAL** Identify the properties of a 2-D shape that stay the same after a rotation.

Rotation 1

Trapezoid *ABCD* has been rotated counterclockwise about point *O* two times. Point *O* is the **centre of rotation**.

Follow these instructions to explore the properties of a rotation.

A. Use a protractor to measure the angles of the rotation. Write the angle measurements on the diagram below.

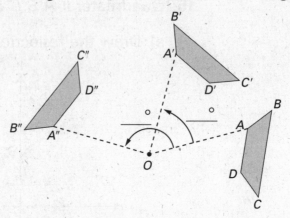

B. Use a compass to draw a circle that has centre *O* and goes through point *B*.
What other points does the circle go through?

_____

Why do you think the circle goes through these points as well?

_____

_____

Rotation 2

Use these steps to rotate point *A* 90° clockwise about the origin, *O*.

Step 1: Join point *A* to the origin, *O*, with a dashed line, as shown.

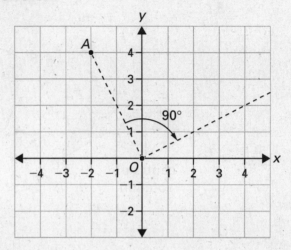

MATH TERM

clockwise

Step 2: Using a protractor, mark a 90° angle **clockwise** from the line. Draw a dashed line from *O* to your mark to form the 90° angle, as shown.

Step 3: Place the point of your compass on *O* and the pencil tip on *A*. Draw an arc from *A* to the dashed line created in Step 2.

Step 4: Make a point where the arc meets the dashed line and label it *A'*.
Name the coordinates of *A* and *A'*.

$$A(\underline{\hspace{1cm}}, \underline{\hspace{1cm}}) \rightarrow A'(\underline{\hspace{1cm}}, \underline{\hspace{1cm}})$$

Use this step to rotate point *A* 90° counterclockwise about the origin, *O*, and create image point *A"*.

MATH TERM

counterclockwise

Step 5: Repeat Steps 2 to 4, except in Step 2, mark a 90° angle **counterclockwise** from the line. Label the new point *A"*.
Name the coordinates of *A"*. $A"(\underline{\hspace{1cm}}, \underline{\hspace{1cm}})$

Reflecting

▶ Look at the trapezoid in Rotation 1.
  (Circle) the property that changes after a rotation.

  orientation   shape   size

TURN

Practising

Text page 242

4. Pentagon *ABCDE* is rotated about the centre of rotation. (Circle) the shapes that could be images of *ABCDE* after a rotation.

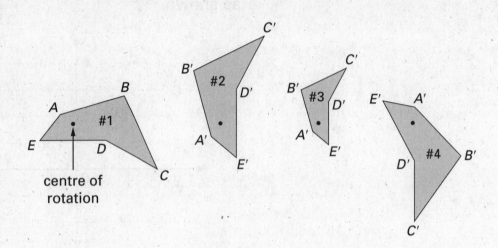

Shape # _____ *cannot* be a rotated image of

pentagon *ABCDE* because _____

_____

8. Rotate parallelogram *DEFG* 90° clockwise about the origin. Point *D* has been rotated for you.
Label the vertices of the image *D′E′F′G′*.

Hint

Rotate each point separately. Then join the points to form the parallelogram.

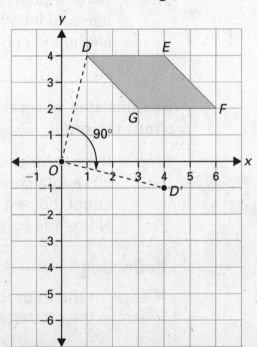

15. The vertices of △*DEF* have coordinates *D*(−4, 4), *E*(−1, 4), and *F*(0, 1).

a) Plot △*DEF* on the grid.

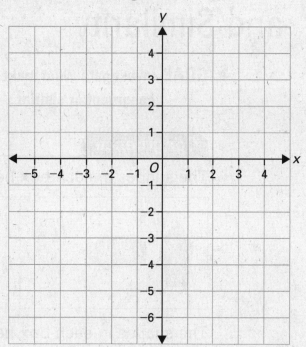

b) Rotate △*DEF* 180° counterclockwise about the origin, *O*. Label the image *D′E′F′*.

c) Name the coordinates of the image.

D′(_____, _____)

E′(_____, _____)

F′(_____, _____)

Congruence and Similarity

▶ **GOAL** Investigate what makes two shapes congruent or similar.

Congruence

The shapes in each box below are congruent.

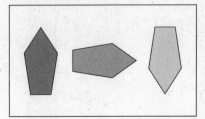 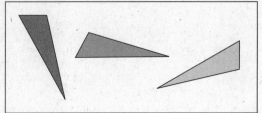

The shapes in each box below are *not* congruent.

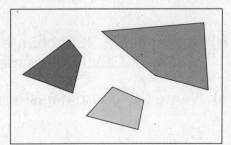 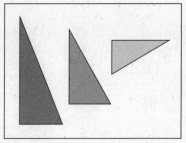

Cut out the shapes and ◣ from Cutout Page 7.5.

How can you use the cutouts to check whether the shapes in the first two boxes are congruent?

_____

_____

_____

Similarity

The shapes in each box below are similar.

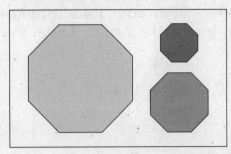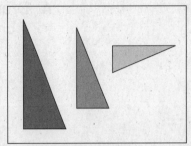

The shapes in the box below are *not* similar.

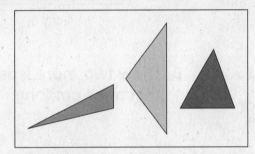

Cut out the shapes and ◢ from Cutout Page 7.5.

Match one of the angles on the cutout to an angle on the similar shape above. What do you notice about the sides?

_____

Connect Your Work

Use the diagrams on page 160 to describe what you think "congruent" means.

_____

_____

_____

> **Hint**
> Include angles and side lengths in your descriptions.

Use the diagrams above to describe what you think "similar" means.

_____

_____

_____

Follow these instructions to explore congruent and similar triangles.

A. Draw another equilateral triangle with sides that are 4 cm long. Use a ruler. Label the side lengths.

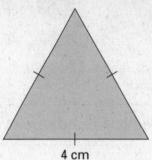

4 cm

B. Draw two more isosceles triangles with equal sides that are 4 cm long. Use a ruler. Label the side lengths.

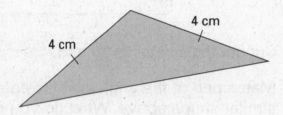

4 cm   4 cm

C. Draw two more triangles with angles of 30°, 80°, and 70°. Use a ruler and a protractor. Label the angles.

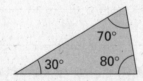

70°   30°   80°

D. Draw two more triangles with a 5 cm side between angles of 60° and 70°. Use a ruler and a protractor. Label the 5 cm side length and the two angles.

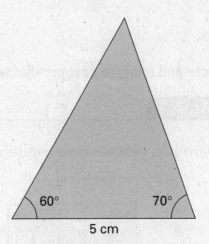

60°   70°

5 cm

Complete the table using the triangles you drew.

Group of triangles	How are the triangles the same?	How are the triangles different?	Are the triangles congruent?	Are the triangles similar?
A. Equilateral triangles with sides that are 4 cm				
B. Isosceles triangles with equal sides that are 4 cm				
C. Triangles with angles of 30°, 80°, and 70°				
D. Triangles with a 5 cm side between angles of 60° and 70°				

Reflecting

▶ Is it easier to tell whether two shapes are congruent or whether they are similar? Explain.

_____

_____

Tessellations

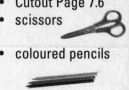
▶ **GOAL** Create and analyze designs that tessellate a plane.

Tessellation 1

MATH TERM

tessellation
an arrangement of
shapes that are the
same shape and size
and cover an area
with no gaps or
overlaps

T pentomino

This diagram is a **tessellation** made with the T pentomino.

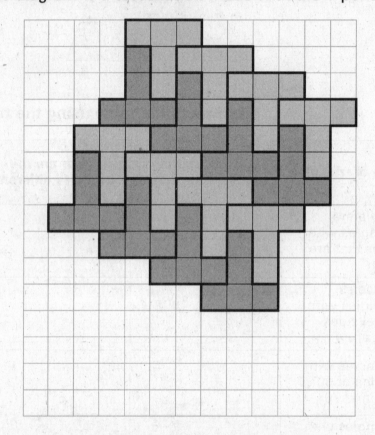

Use these steps to continue the tessellation.

Step 1: Cut out the T pentomino from Cutout Page 7.6.

Step 2: Use the cutout to continue the tessellation.
Outline the shapes. Use coloured pencils to
shade the two orientations.

Step 3: Write the transformation used to move from
T to ⊥.

Hint

The two orientations
are:

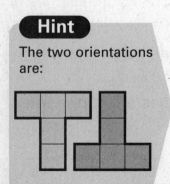

L pentomino

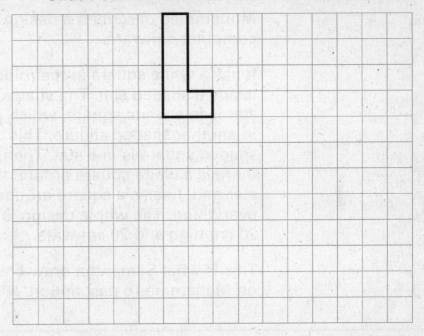

Use these steps to create a tessellation for the L pentomino.

Step 1: Cut out the L pentomino from Cutout Page 7.6.

Step 2: Use the L pentomino to make a tessellation on the grid. The first one is drawn for you. Shade it green. Shade each orientation of L in a different colour.

Step 3: Complete the table to describe how you would transform the green L into the other orientations.

Colour of orientation	Rotation used

Reflecting

▶ Would the T pentomino and/or the L pentomino be good shapes to tile a kitchen floor? Explain.

_____

_____

_____

Communicating about Geometric Patterns

▶ **GOAL** Describe designs in terms of congruent, similar, and transformed images.

Mohammed described a design that he made with computer software:

It has a white square in the middle, inside a shaded star. The shaded star is in a white square, which is in another shaded square. This shaded square is in a star. Then, there is a white square around the star and, finally, a square around everything. The whole design is 20 cm high and 20 cm wide.

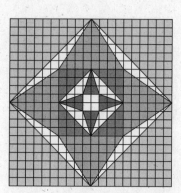

This is what Samantha drew based on Mohammed's description. ▶

A. Compare Samantha's design to the original. What parts of her design match the original?

_____

_____

_____

_____

_____

B. What parts of Samantha's design are different from the original?

_____

_____

_____

_____

C. <u>Underline</u> the parts of Mohammed's description that need more detail.

D. Here are some questions you could ask Mohammed to help him improve his description:

- Did you completely describe both the shape and orientation of each part of your design?
- Did you describe all the transformations you used?
- Did you describe any equal sides and use appropriate measurements?

Suggest specific ways Mohammed could improve his description. Use the questions above and the Communication Checklist for help.

_____

_____

_____

_____

_____

_____

Communication Checklist

❑ Was your description clear?

❑ Was your description complete and thorough?

❑ Did you use necessary and appropriate math language?

Reflecting

▶ Which questions in the Communication Checklist are covered in the questions in Part D?

_____

_____

TURN ▶

Practising

Text page 252

3. Consider the following design.
Assume each square in the grid is 1 cm by 1 cm.

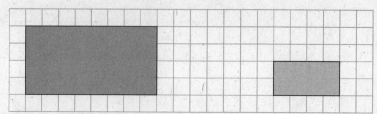

Communication Checklist

❑ Was your description clear?

❑ Was your description complete and thorough?

❑ Did you use necessary and appropriate math language?

a) Describe the design. Keep the Communication Checklist in mind and answer these questions in your description:

• What are the dimensions of the grid?

• What are the dimensions of the rectangles?

• Where are the rectangles located on the grid?

_____

_____

_____

_____

_____

_____

_____

_____

_____

_____

b) Ask someone else to use your description to draw the design. Consider how closely the designs match. Add any details to improve your description.

4. This design is made up of four triangles.
Assume each square in the grid is 1 cm by 1 cm.

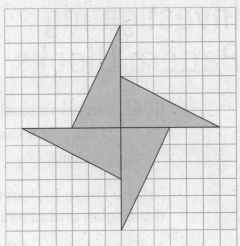

a) Describe the design. Keep the Communication
Checklist in mind and answer these questions in
your description:

- What are the dimensions of the grid?

- How many triangles are there?

- Where are the triangles located on the grid?

- What kind of triangles are they?

- What are the dimensions of the triangles?

_____

_____

_____

_____

_____

_____

_____

_____

b) Ask someone else to use your description to draw
the design. Consider how closely the designs
match. Add any details to improve your
description.

Lesson 7.7: Communicating about Geometric Patterns **169**

END

7.8
Text page 254

Investigating Pattern Blocks

You will need
- Cutout Page 7.8
- pattern blocks
- paper
- scissors
- a calculator
- a protractor

▶ **GOAL** Use transformations and properties of congruent shapes to solve problems.

Follow these instructions to explore the angles and areas of each pattern block.

A. Cut out each of these blocks from Cutout Page 7.8.

B. Fold the cutouts to determine which of the pattern blocks are **regular polygons**. Fold sides onto sides and angles onto angles.
Complete the second column of the table.

MATH TERM

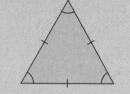

regular polygon
a polygon with all sides equal and all angles equal

Pattern block	Regular polygon (Yes/No)	Fraction of hexagon	Number of blocks to form 360°	Vertex angle measure
yellow hexagon			3	$\frac{360°}{3} =$
red trapezoid				
blue rhombus				
orange square				
green triangle				
beige rhombus				

170 Lesson 7.8: Investigating Pattern Blocks

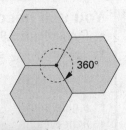

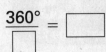

C. Three hexagons form a tessellation around a single point. The angle formed where the three hexagons meet is 360°.

Divide 360° by the number of angles that form 360° to find the measure of each vertex angle of a regular hexagon.

$$\frac{360°}{\Box} = \Box$$

D. Fit each type of pattern block into a tessellation around a single point.

E. Calculate the vertex angles of each block. (See Part C.) Complete the last two columns of the table. Use a protractor to check the angle measures.

F. Cover the hexagon completely with green triangles. How many triangles did it take? _____

$$\text{green triangle} = \frac{1}{\Box} \text{ of a yellow hexagon}$$

G. Repeat Part F by covering the hexagon completely with red trapezoids. Then repeat Part F again with the blue rhombus. Complete the shaded, third column of the table.

H. Try to cover the hexagon with the orange square and then the beige rhombus. What do you notice?

_____

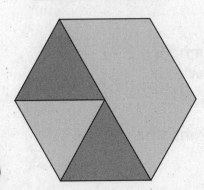

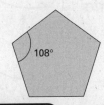

Reflecting

▶ A regular pentagon has a vertex angle of 108°. Cut out the pentagon from Cutout Page 7.8 and try to form a tessellation. Explain why a regular pentagon will not tessellate.

_____

_____

_____

_____

Tessellating Designs

▶ **GOAL** Create irregular tiles, and determine whether irregular tiles can be used to tessellate a plane.

Use these steps to create an irregular tile and then tessellate a plane with the tile.

Hint

You do not have to copy the design in the diagrams. Try making up your own design.

Step 1: Cut out the larger square with the dashed outline on Cutout Page 7.9.

Step 2: Change the left side of the smaller, inner square on the cutout with a curve.

Step 3: Translate the curve to the right side of the square.

Step 4: Change the bottom of the square with a different curve.

Step 5: Translate the curve to the top of the square.

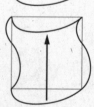

Step 6: Cut out your tile and glue it to light-weight cardboard. Cut off the extra cardboard around your tile.

Step 7: Trace around your tile to make a tessellation in the box on page 173.

Reflecting

▶ Would your tile tessellate if you drew a different curve on each of the sides? How do you know?

_____

_____

_____

_____

_____

END

Exploring Pattern Representations

▶ **GOAL** Explore different ways of describing a pattern.

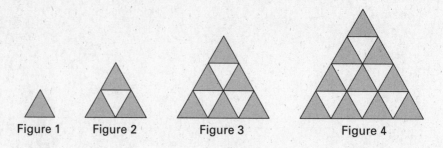

Figure 1   Figure 2   Figure 3   Figure 4

A. There are many different patterns in the triangle pattern above. Complete the table to show the number sequence for each pattern.

Pattern	Figure 1	Figure 2	Figure 3	Figure 4
number of rows	1			
number of small triangles	1			
number of grey triangles	1			
number of white triangles	0			
side length of figure	1			

Hint

Count the side of a small triangle as one unit.

B. Choose one of the patterns in Part A.

_____

Hint

The Term value is the number of rows, number of triangles, or side length for a figure in the pattern you chose.

C. Complete the table of values for the pattern you chose.

Figure number	Term value
1	
2	
3	
4	
5	
6	

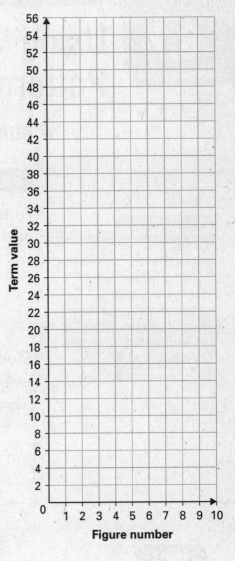

D. Use the table of values to complete the scatter plot for the pattern.

E. Use your graph to determine the Term value for Figure 10.

_____

F. Write the pattern rule.

_____

_____

Reflecting

▶ Circle the representation that you would use to determine the term value of Figure 57.

table of values   scatter plot   pattern rule

Why would you choose this representation?

_____

_____

Using Variables to Write Pattern Rules

▶ **GOAL** Use a variable to write a pattern rule.

Pattern 1

MATH TERM

variable
a letter or symbol,
such as *n*, that
represents a number;
it can represent the
quantity that changes
in a pattern

Follow these steps to write a pattern rule for this
pattern using a **variable**.

Figure 1   Figure 2   Figure 3   Figure 4

Step 1: Complete the table using the pattern above.

Figure number	Number of rectangles	Number of triangles	Total number of shapes
1			
2			
3			
4			

Hint
The quantity that
stays the same is the
constant.

Step 2: (Circle) what stays the same in the pattern.

number of triangles   number of rectangles

Step 3: How does the Figure number relate to the
number of rectangles?

_____

Step 4: The pattern rule that relates the Figure number
to the *total* number of shapes is

To get the number of shapes, add 1 to the
Figure number.

MATH TERM

algebraic expression
includes one or more
variables and may
include numbers
(constants) and
operation signs

This pattern rule can be written as an **algebraic
expression**, using the variable *n* to represent the
Figure number.

$$n + 1$$

Follow these steps to write a pattern rule for this pattern using a variable.

Figure 1   Figure 2   Figure 3   Figure 4

Step 1: Complete the table using the pattern above.

Figure number	Number of rectangles	Number of triangles	Total number of shapes
1			
2			
3			
4			

Step 2: What stays the same in the pattern?

_____

Step 3: What do you multiply the Figure number by to get the number of rectangles used? _____

Step 4: Write the pattern rule that relates the Figure number to the total number of shapes.

_____

_____

Step 5: Write the pattern rule as an algebraic expression. Let the variable *n* represent the Figure number.

$2n +$ _____
(constant from Step 2)

Hint

$2n$ is the same as $2 \times n$.

Reflecting

▶ Why would you use an algebraic expression, instead of words, to represent a pattern rule?

_____

_____

Practising

Text page 276 **4.**

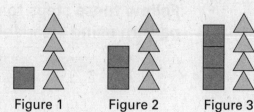

Figure 1   Figure 2   Figure 3

a) (Circle) what stays the same in the pattern above.

number of triangles   number of squares

b) How does the Figure number relate to the number of squares?

_____

Hint

Describe the relationship between the Figure number and the total number of shapes.

c) Write the pattern rule for the total number of shapes using words and numbers.

_____

d) Write the pattern rule as an algebraic expression. Let *n* represent the Figure number.

$n +$ _____

5.

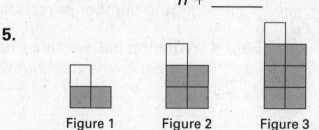

Figure 1   Figure 2   Figure 3

a) (Circle) what stays the same in the pattern above.

number of white squares   number of grey squares

b) What would you multiply the Figure number by to get the number of grey squares? _____

c) Write the pattern rule for the total number of squares using words and numbers.

_____

Hint

Let *n* represent the Figure number.

d) Write the pattern rule as an algebraic expression.

_____ $n +$ _____

7. Two students wrote different algebraic expressions for the same pattern. Explain why each student is correct.

a) $n + 1 + n$

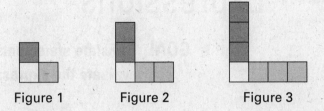

Figure 1   Figure 2   Figure 3

This student is correct because _____

_____

_____

_____

_____

b) $2n + 1$

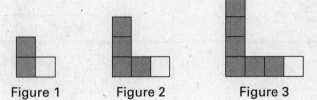

Figure 1   Figure 2   Figure 3

This student is correct because _____

_____

_____

_____

_____

_____

END

Creating and Evaluating Expressions

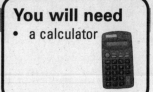

▶ **GOAL** Translate statements into algebraic expressions, then evaluate the expressions.

Angela can choose from two methods of payment to ski at a local ski club:

Plan A: $35 per day for each day she skis

Plan B: $200 for the season plus an additional $15 per day

Angela plans to ski on 8 different days throughout the season. Which plan should she choose?

Use these steps to solve the problem.

Step 1: Complete the tables below.

Plan A		Plan B	
Number of days	**Cost ($)** $35 \times$ number of days	**Number of days**	**Cost ($)** $200 + 15 \times$ number of days
1	35×1 $= 35$	1	$200 + 15 \times 1$ $= 200 + 15$ $= 215$
2	35×2 $= 70$	2	$200 + 15 \times 2$ $= 200 + 30$ $= \underline{\hspace{1cm}}$
3	$35 \times \underline{\hspace{1cm}}$ $= \underline{\hspace{1cm}}$	3	$200 + 15 \times \underline{\hspace{1cm}}$ $= 200 + \underline{\hspace{1cm}}$ $= \underline{\hspace{1cm}}$

Step 2: Write algebraic expressions for each plan. Use the variable *d* to represent the number of days skiing. The algebraic expression for Plan A is done for you.

Hint

Remember, 35*d* is the same as 35 × *d*.

Plan A	Plan B
35*d*	_____ + _____

Step 3: Angela plans to ski on 8 different days. Substitute the number of days for *d* in your algebraic expressions for Plan A and Plan B. Plan A is done for you.

Plan A		Plan B	
Expression:	35*d*	Expression:	_____ + _____
Substitute 8 days:	= 35 × 8	Substitute 8 days:	
Cost for Plan A:	= 280	Cost for Plan B:	

Step 4: Which plan should Angela choose? Why?

_____

_____

Reflecting

▶ Explain why using an algebraic expression to evaluate might be easier than a scatter plot or a table of values.

_____

_____

_____

_____

Practising

Text page 280

Hint

Remember,
4*b* is the same as
4 × *b*.

4. A bowl of chili costs $4. (Circle) the expression that represents the cost of buying chili for *b* people.

$$7b - 4 \qquad b + 4 \qquad 4b$$

5. Evaluate each expression when *d* = 5. The first one is done for you.

a) $6d$
$= 6 \times 5$
$= 30$

b) $d + 1$
$= \underline{\quad} + 1$
$= \underline{\quad}$

c) $5d - 1$
$= 5 \times \underline{\quad} - 1$
$= \underline{\quad} - 1$
$= \underline{\quad}$

6. The cost of renting a sleigh is $12 per hour plus $35 for the driver.

a) (Circle) the fee that stays the same no matter how many hours the sleigh is rented for.

$12 per hour $35 for the driver

b) Write an algebraic expression to represent the cost of renting the sleigh. Use *h* to represent the number of hours.

$\underline{\quad} + \underline{\quad}$

8. Evaluate each algebraic expression when *a* = 3. The first one is done for you.

a) $3a$
$= 3 \times 3$
$= 9$

b) $8a$
$=$
$=$

c) $3a + 2$
$=$
$=$
$=$

d) $4a - 6$
$=$
$=$
$=$

10. Write an algebraic expression for each cost. The first one is done for you.

a) Hamburgers cost $3 each.
Let h be the number of hamburgers sold.

$3h$

b) Hats are on sale for $10 each.
Let h be the number of hats sold.

_____

c) Renting skates costs $2 per hour plus $5.
Let h be the number of hours.

_____

12. Jerry sells toques at a kiosk.
He is paid $25 a day plus $2 for each toque he sells.

a) What part of his salary does not change, no matter how many toques he sells?

_____

b) Write an algebraic expression that describes Jerry's salary. Choose a variable to represent the number of toques he sells.

_____

Hint
Choose an appropriate letter for the variable.

c) Use your algebraic expression to calculate how much Jerry will earn in one day if he sells 17 toques.

Lesson 8.3: Creating and Evaluating Expressions **183**

END

Solving Equations by Inspection

You will need
- a calculator
- toothpicks (optional)

▶ **GOAL** Write equations and solve them by inspection.

John and Heidi are building this toothpick pattern. They have 28 toothpicks.

Figure 1   Figure 2   Figure 3

Use these steps to determine how many squares John and Heidi can build for this pattern with 28 toothpicks.

> **Hint**
>
> You can model Figure 4 with toothpicks.

Step 1: Complete the table of values.

Step 2: Write the pattern rule for the relationship between the number of squares and the number of toothpicks.

Number of squares	Number of toothpicks
1	
2	
3	
4	

_____

_____

Step 3: Write an algebraic expression for the pattern rule. Let *s* represent the number of squares.

_____$s +$ _____

Step 4: If *t* represents the total number of toothpicks you need, complete the **equation**.

$t =$ _____$s +$ _____

> **MATH TERM**
>
> **equation**
> a statement in which the value on the left side of the equal sign is the same as the value on the right side of the equal sign

Step 5: Since John and Heidi have 28 toothpicks, rewrite the equation substituting 28 for *t*.

_____ $=$ _____$s +$ _____

Step 6: Use guessing and testing to find what 3*s* represents. This step is partly done for you.

$$28 = 3s + 1$$
$$28 = 27 + 1$$
So, $3s = 27$.

What number $\times$ 3 = 27? _____
So, s = _____

This method is called solving by inspection.

How many squares can they build? _____

Hint

Remember,
s = number of squares.

Step 7: Check your solution.

Left side	Right side
28	$3s + 1$
	$= 3 \times$ _____ $+ 1$
	$=$ _____ $+ 1$
	$=$ _____

Reflecting

▶ Why is solving by inspection a better way to solve this problem than continuing the table of values?

_____

_____

_____

Practising

Text page 286

3. Solve each of the following equations by inspection.

a) $w - 11 = 22$
If you subtract 11 from a number, the result is 22.
The number is _____, since _____ $- 11 = 22$.

So, w = _____

b) $9p = 63$
If you multiply a number by 9, the result is 63.
The number is _____, since $9 \times$ _____ $= 63$.

So, p = _____

4. Rana wrote the expression $4n + 3$ to describe the pattern below, when n is the Figure number.

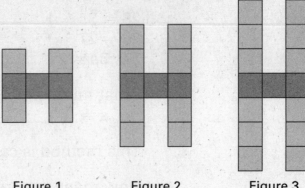

Figure 1   Figure 2   Figure 3

a) Why did Rana add 3 to the expression?

_____

b) Why did Rana multiply the Figure number by 4?

_____

c) The following equation describes the number of squares needed for each figure in this pattern. Let s represent the number of squares.

$$s = 4n + 3$$

Rana built a figure in this pattern that has 23 squares. She substituted 23 for s in the equation.

$$23 = 4n + 3$$

If you add 3 to $4n$, the result is 23. What number does $4n$ represent?

$$4n = \underline{\hspace{1cm}}$$

So, $n = \underline{\hspace{1cm}}$, since $4 \times \underline{\hspace{1cm}} = 20$.

Hint

Divide the number $4n$ by 4 to find n.

d) Check your solution.

Left side	Right side
23	$4n + 3$
	$= 4 \times \underline{\hspace{1cm}} + 3$
	$= \underline{\hspace{1cm}} + 3$
	$= \underline{\hspace{1cm}}$

5. Solve each equation by inspection.

a) $7b = 84$

   $7 \times$ _____ $= 84$

   So, $b =$ _____

b) $8 + z = 30$

   $8 +$ _____ $= 30$

   So, $z =$ _____

c) $2w + 1 = 17$

   _____ $+ 1 = 17$

   So, $2w =$ _____

         $w =$ _____

d) $9n - 4 = 32$

   _____ $- 4 = 32$

   So, $9n =$ _____

        $n =$ _____

Text page 287   **9.**

Figure 1   Figure 2   Figure 3

a) Write an algebraic expression for the total number of shapes in the pattern. Use *n* to represent the Figure number.

_____$n +$ _____

b) Write the equation that you can solve to determine the Figure number that has 28 shapes.

$28 =$ _____$n +$ _____

c) Solve the equation. Show your work.

$28 =$ _____$n +$ _____
$28 =$ _____ $+$ _____

So, $2n =$ _____
$n =$ _____

d) Check your solution.

Left side   Right side
28   _____$n +$ _____
           $=$ _____ $\times$ _____ $+$ _____
           $=$ _____ $+$ _____
           $=$ _____

8.5 Solving Equations by Systematic Trial

You will need
• a calculator

▶ **GOAL** Write equations and solve them using guessing and testing.

The first three figures of a pattern are shown below.

Figure 1   Figure 2   Figure 3

Use these steps to determine which Figure number has 33 symbols.

Step 1: Extend the pattern above to complete the table.

Figure number	Number of happy faces	Number of suns	Total number of symbols
1	2	3	5
2	4	3	7
3	6		
4			

Step 2: (Circle) what stays the same in the pattern.

number of suns   number of happy faces

Step 3: What do you have to multiply the Figure number by to get the number of happy faces? _____

Step 4: Write an equation that you would use to find the total number of symbols (*t*) in each figure.
Let *n* be the Figure number.

$$t = \underline{\quad\quad} n + \underline{\quad\quad}$$

Step 5: Write the equation that you would use to determine which figure requires 33 symbols.

_____ = _____ n + _____

Step 6: Use systematic guessing and testing to solve the equation.

- First predict what n is.
- Then evaluate the equation substituting your prediction for n.
- Then write whether your prediction was too high, too low, or correct. This will help you decide what number to choose next.

One prediction is made for you.
The prediction is too low, so your answer must be greater than 10. Try two more predictions.

Systematic Trial		
Predict n	**Evaluate $2n + 3$**	**Too high? Too low? Correct?**
10	$2 \times 10 + 3$ $= 20 + 3$ $= 23$	too low

Reflecting

▶ When using systematic trial, why is it not necessary to check your solution?

_____

_____

_____

Practising

Text page 290 **4.** (Circle) the correct answer on the right.

    a) $4r = 16$
        $r = 4$ OR $r = 64$

    Explain your thinking.

_____

    b) $p - 9 = 15$
        $p = 6$ OR $p = 24$

    Explain your thinking.

_____

5. Solve each equation by completing the table.

    a) $n - 12 = 122$

Predict n	Evaluate $n - 12$	Too high? Too low? Correct?
144	$144 - 12$ $= 132$	too high
124		
134		

    b) $15b = 315$

Predict b	Evaluate $15b$	Too high? Too low? Correct?
15	15×15 $=$	
25		
21		

8. Use systematic trial to find the value of each variable.

a) $9r = 792$

Predict r	Evaluate $9r$	Too high? Too low? Correct?

b) $5 + 4c = 105$

Predict c	Evaluate $5 + 4c$	Too high? Too low? Correct?

END

Communicating the Solution for an Equation

▶ **GOAL** Communicate about solving equations.

Anthony says he knows how many counters are in the container on this pan balance. He tries to show his thinking using algebra.

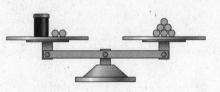

Anthony's Explanation

$c + 2 = 6$
$c + 2 - 2 = 6 - 2$
$c = 4$

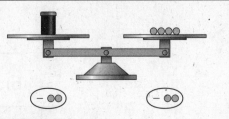

There are 4 counters in the container.

Kaitlyn asks Anthony to explain what he means.

Kaitlyn's Questions

1. What does c represent?

2. Why did you subtract 2 from each side?

3. How do you know your solution is correct?

A. Which parts of Anthony's explanation were not clear? Use the Communication Checklist for help.

_____

_____

_____

B. Would you have asked the same questions as Kaitlyn? Explain why or why not.

_____

_____

Communication Checklist

❑ Did you show each step of your thinking?

❑ Did you express yourself clearly?

❑ Were you convincing?

Anthony read Kaitlyn's questions and then rewrote his explanation.

Anthony's New Explanation

I wrote the balance problem as an algebraic equation. I let c represent the number of counters in the container.

$c + 2 = 6$

To find out how many counters are in just the container, I subtracted 2 from the left side.

To balance the right side with the left side, I subtracted 2 from the right side as well.

$c + 2 - 2 = 6 - 2$
$c = 4$

I checked my solution by substituting 4 for the variable and checking that it made the equation true.

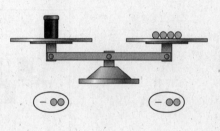

Left side	Right side
$c + 2$	6
$= 4 + 2$	
$= 6$	

There are 4 counters in the container.

Reflecting

► Why is it important to show your steps in a logical order when solving an algebraic equation?

_____

_____

_____

_____

_____

TURN

Practising

Text page 293 **4.** Explain what is happening in each step of Tynessa's solution.

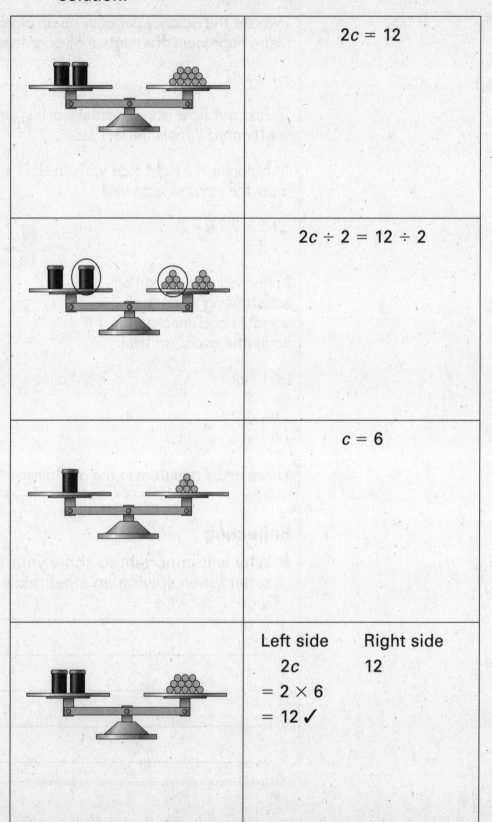

| | $2c = 12$ |

| | $2c \div 2 = 12 \div 2$ |

| | $c = 6$ |

| | Left side   Right side
   $2c$ 12
$= 2 \times 6$
$= 12$ ✓ |

Text page 294

6. Use pictures and words to explain each step in the solution to this balance problem. The first step is done for you. Use the Communication Checklist to write a clear solution.

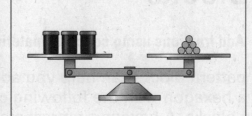	**Step 1:** An equation for this pan balance situation is $3c = 6$.
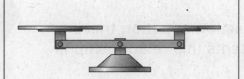	**Step 2:**
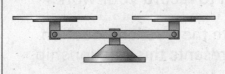	**Step 3:**
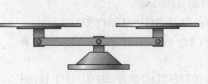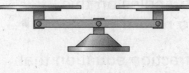	**Step 4:**

9.1

Text page 306

Adding Fractions with Pattern Blocks

You will need
- pattern blocks

▶ **GOAL** Add fractions using concrete materials.

Using pattern blocks can help you add fractions. Cover a hexagon with the following combinations of pattern blocks. A hexagon represents 1 whole.

Hint

Each pattern block represents the following values.

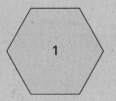

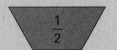

Combination 1

Cover the yellow hexagon with red trapezoids.

The fraction equation that represents this relationship is

$\frac{1}{2} + \frac{1}{2} = 1$.

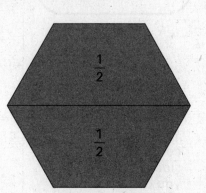

Combination 2

Cover the yellow hexagon with blue rhombuses.
Trace the rhombuses on this hexagon to record your work.▶

Complete the fraction equation that represents this relationship.

$\frac{1}{3} + \underline{\quad} + \underline{\quad} = 1$

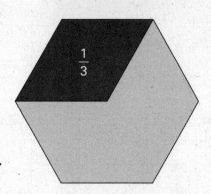

Combination 3

Cover the yellow hexagon with green triangles.
Trace the triangles on this hexagon to record your work.▶

Write the fraction equation that represents this relationship.

$\underline{\quad} + \underline{\quad} + \underline{\quad} + \underline{\quad} + \underline{\quad} + \underline{\quad} = 1$

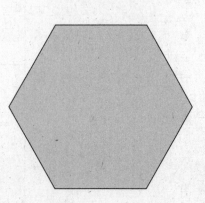

196 Lesson 9.1: Adding Fractions with Pattern Blocks

Copyright © 2006 Nelson

Mix the types of pattern blocks to find all the possible ways to cover the yellow hexagon. Trace the pattern blocks on the hexagons below to record your work. Write equations that represent the relationships. There may be more hexagons than you need.

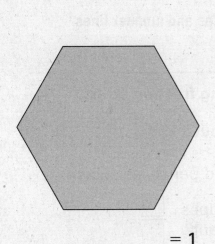

_____ = 1

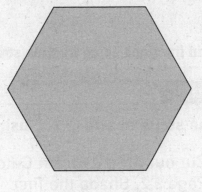

_____ = 1

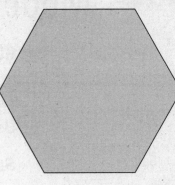

_____ = 1

_____ = 1

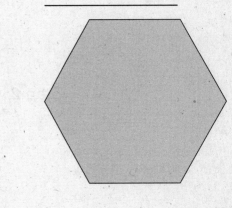

_____ = 1

Hint

Be sure to count the combinations on page 196.

How many ways, in total, was the yellow hexagon covered? _____

Explain how you know that there are no more ways.

_____

_____

Reflecting

▶ Explain why $\frac{1}{2} + \frac{1}{2} = 1$ can be written as $\frac{1}{2} + \frac{1}{2} = \frac{2}{2}$.

_____

_____

_____

Adding Fractions with Models

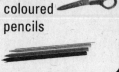

▶ **GOAL** Add fractions using fraction strips and number lines.

Method 1

Use these steps to add $\frac{1}{4} + \frac{1}{3}$ using fraction strips.

Step 1: Cut out Strip #1 from Cutout Page 9.2. Shade the first section with a red coloured pencil, as shown.

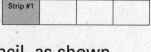

What fraction of the strip does the shaded section represent? ☐/☐

Step 2: Cut out Strip #2 from Cutout Page 9.2. Shade the first section with a blue coloured pencil, as shown.

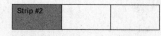

What fraction of the strip does the shaded section represent? ☐/☐

Step 3: Cut out Strip #3 from Cutout Page 9.2.
Line up Strip #1 and Strip #3.
How many sections on Strip #3 are equivalent to one red $\frac{1}{4}$ section on Strip #1? _____
Shade this number of sections red on Strip #3.

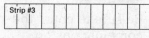

Complete. $\frac{1}{4} = \frac{\square}{12}$

Step 4: Line up Strip #2 and Strip #3.
How many sections on Strip #3 are equivalent to one blue $\frac{1}{3}$ section on Strip #2? _____
Shade this number of sections blue on Strip #3.
Start shading after the red sections.

Complete. $\frac{1}{3} = \frac{\square}{12}$

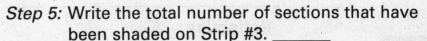

Step 5: Write the total number of sections that have been shaded on Strip #3. _____

Write this as a fraction of the whole strip.

Hint

The denominator is the total number of sections on the fraction strip.

Step 6: Complete. $\frac{1}{4} + \frac{1}{3} = \frac{\square}{12} + \frac{\square}{12}$

$$= \frac{\square}{12}$$

Method 2

Use these steps to add $\frac{1}{4} + \frac{1}{3}$ on a number line.

The number line below starts at 0 and ends at 1. It has been divided into 12 sections because 12 is a **common denominator** for $\frac{1}{3}$ and $\frac{1}{4}$.

MATH TERM

common denominator
a common multiple of the denominators of two or more fractions

Step 1: Start at 0 and draw an arrow to show where $\frac{1}{4}$ would end on the number line. This step is done for you. (Remember, $\frac{1}{4} = \frac{3}{12}$.)

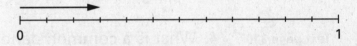

Step 2: Draw a second arrow to show where $\frac{1}{3}$ would end on the number line. (Remember, $\frac{1}{3} = \frac{4}{12}$.)

Step 3: To show $\frac{1}{4} + \frac{1}{3}$, redraw the $\frac{1}{3}$ arrow but start it at the end of the $\frac{1}{4}$ arrow. This step is done for you.

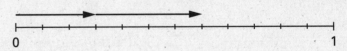

Count the total number of twelfths. _____

So, $\frac{1}{4} + \frac{1}{3} = \frac{\square}{12}$

Reflecting

▶ Change $\frac{1}{4}$ and $\frac{1}{3}$ so they each have a denominator of 12.

$$\frac{1}{4} = \frac{1 \times \square}{4 \times \square}$$
$$= \frac{\square}{12}$$

$$\frac{1}{3} = \frac{1 \times \square}{3 \times \square}$$
$$= \frac{\square}{12}$$

▶ 12 is a common denominator for $\frac{1}{4}$ and $\frac{1}{3}$. Why are common denominators important for adding fractions?

_____

_____

_____

_____

Practising

Text page 310

4. What is a common denominator for $\frac{3}{4}$ and $\frac{1}{6}$? _____

Change $\frac{3}{4}$ and $\frac{1}{6}$ so they each have a denominator of 12.

$$\frac{3}{4} = \frac{3 \times \square}{4 \times \square}$$
$$= \frac{\square}{12}$$

$$\frac{1}{6} = \frac{1 \times \square}{6 \times \square}$$
$$= \frac{\square}{12}$$

Shade the fraction strip to add $\frac{3}{4} + \frac{1}{6} = \frac{\square}{\square}$.

7. Divide the fraction strip into equal sections to represent a common denominator for $\frac{1}{4}$ and $\frac{1}{2}$.

Shade the fraction strip above to add $\frac{1}{4} + \frac{1}{2} = \frac{\Box}{\Box}$.

8. a) What is a common denominator for $\frac{3}{5}$ and $\frac{1}{4}$?
_____

Change $\frac{3}{5}$ and $\frac{1}{4}$ so they each have the common denominator.

$$\frac{3}{5} = \frac{3 \times \Box}{5 \times \Box} \qquad\qquad \frac{1}{4} = \frac{1 \times \Box}{4 \times \Box}$$

$$= \frac{\Box}{\Box} \qquad\qquad\qquad = \frac{\Box}{\Box}$$

How many sections should there be on the number line? _____

0 ——————————————————————— 1

Use the number line to add $\frac{3}{5} + \frac{1}{4} = \frac{\Box}{\Box}$.

b) Change $\frac{2}{3}$ and $\frac{1}{6}$ so they each have a common denominator.

$$\frac{2}{3} = \frac{2 \times \Box}{3 \times \Box} \qquad\qquad \frac{1}{6} = \frac{1 \times \Box}{6 \times \Box}$$

$$= \frac{\Box}{\Box} \qquad\qquad\qquad = \frac{\Box}{\Box}$$

0 ——————————————————————— 1

Use the number line to add $\frac{2}{3} + \frac{1}{6} = \frac{\Box}{\Box}$.

END

Multiplying a Whole Number by a Fraction

You will need
- counters

▶ **GOAL** Multiply fractions by whole numbers.

Method 1

3 packages of gum each have $\frac{5}{8}$ left. How many full packages are there?

Follow these instructions to multiply $3 \times \frac{5}{8}$ **using grids.**

A. Each grid below represents 1 package containing 8 pieces. Place counters on each grid to model 3 packages with 5 of the 8 pieces ($\frac{5}{8}$) in each of them.

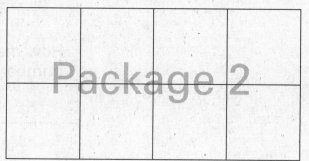

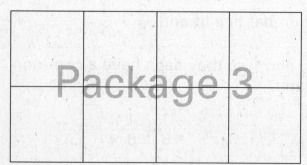

How many counters did you use in total? _____

This model represents $3 \times \frac{5}{8} = \frac{15}{8}$.

B. Remove the 15 counters. Then replace the 15 counters on the grid. This time, completely fill one grid before moving to the next grid.

How many grids did you fill completely? _____
How many counters are only partly filling a grid? _____

Write these leftover counters as a fraction. $\frac{\square}{\square}$

So, $3 \times \frac{5}{8} = \frac{15}{8}$, or $1\frac{7}{8}$ full packages of gum.

Method 2

Follow these instructions to multiply $4 \times \frac{2}{3}$ **using a number line.**

A. Draw an arrow to represent $\frac{2}{3}$. This step is done for you.

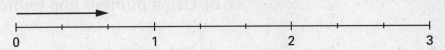

B. How many $\frac{2}{3}$ arrows would you need to represent $4 \times \frac{2}{3}$? _____

C. Draw the 4 arrows end to end starting at 0 to model $4 \times \frac{2}{3}$. The first arrow is shown.

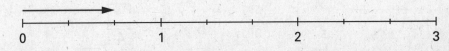

D. Count the number of thirds the arrows represent.

_____

E. What is the end point of the arrows? $\underline{}\dfrac{\square}{3}$

F. Complete. Write the answer as an **improper fraction** and as a **mixed number**.

$$4 \times \frac{2}{3} = \frac{\square}{3}$$

$$= \underline{}\frac{\square}{3}$$

Reflecting

▶ Why can we write $3 \times \frac{5}{8}$ as $\frac{5}{8} + \frac{5}{8} + \frac{5}{8}$?

_____

_____

_____

▶ How do you know that $\frac{7}{3} = 2\frac{1}{3}$?

_____

_____

_____

Practising

6. a) Write $5 \times \frac{3}{4}$ as a repeated addition sentence.

_____

b) Use a number line to model $5 \times \frac{3}{4}$.

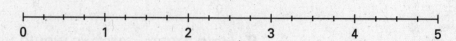

Count the number of fourths the arrows represent. _____

Complete. Write the answer as an improper fraction and as a mixed number.

$5 \times \frac{3}{4} = \dfrac{\boxed{}}{4}$

$= \underline{}\dfrac{\boxed{}}{\boxed{}}$

7. Use counters and the grids below to multiply.

a) $2 \times \frac{1}{3} = \dfrac{\boxed{}}{\boxed{}}$

b) $3 \times \frac{3}{4} = \dfrac{\boxed{}}{4}$

$= \underline{}\dfrac{\boxed{}}{\boxed{}}$

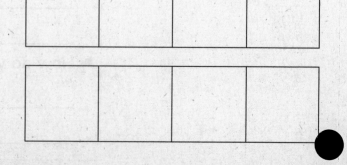

8. a) Write $2 \times \frac{5}{4}$ as a repeated addition sentence.

_____

Use the number line to model $2 \times \frac{5}{4}$.

Why is the number line divided into fourths?

_____

Complete. Write the answer as an improper fraction and as a mixed number.

$2 \times \frac{5}{4} = \dfrac{\square}{\square}$

$= \underline{} \dfrac{\square}{\square}$

b) Write $6 \times \frac{3}{5}$ as a repeated addition sentence.

_____

Use the number line to model $6 \times \frac{3}{5}$.

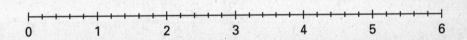

Complete. Write the answer as an improper fraction and as a mixed number.

$6 \times \frac{3}{5} = \dfrac{\square}{\square}$

$= \underline{} \dfrac{\square}{\square}$

END

9.4 Subtracting Fractions with Models

Text page 316

You will need
- a ruler

▶ **GOAL** Subtract fractions using fraction strips and number lines.

Method 1

Use these instructions to estimate and subtract $\frac{3}{4} - \frac{1}{6}$.

A. Use these fraction strips to estimate the difference.

This fraction strip models $\frac{3}{4}$.

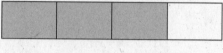

What fraction does this fraction strip model? $\frac{\square}{\square}$

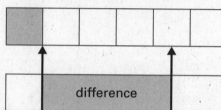

Estimate the difference. _____

difference

B. A common denominator for $\frac{3}{4}$ and $\frac{1}{6}$ is 12.

Change $\frac{3}{4}$ and $\frac{1}{6}$ so they each have 12 as the denominator.

$$\frac{3}{4} = \frac{3 \times \square}{4 \times \square} \qquad \frac{1}{6} = \frac{1 \times \square}{6 \times \square}$$

$$= \frac{\square}{12} \qquad\qquad = \frac{\square}{12}$$

Hint

See Part B to find out how many twelfths $\frac{3}{4}$ and $\frac{1}{6}$ are equivalent to.

C. Shade $\frac{3}{4}$ of this strip.

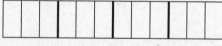

Shade $\frac{1}{6}$ of this strip.

Shade the difference.

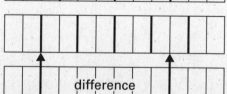

difference

D. Complete. $\dfrac{3}{4} - \dfrac{1}{6} = \dfrac{\square}{12} - \dfrac{\square}{12}$

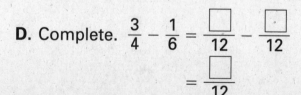

$$= \dfrac{\square}{12}$$

206 Lesson 9.4: Subtracting Fractions with Models

Copyright © 2006 Nelson

Method 2

Follow these instructions to subtract $\frac{2}{3} - \frac{1}{5}$ using a number line.

A. Change $\frac{2}{3}$ and $\frac{1}{5}$ so they each have a common denominator of 15.

$$\frac{2}{3} = \frac{2 \times \square}{3 \times \square}$$

$$= \frac{\square}{15}$$

$$\frac{1}{5} = \frac{1 \times \square}{5 \times \square}$$

$$= \frac{\square}{15}$$

Hint

This number line is divided into 15 sections to match the common denominator.

B. Draw an arrow to represent $\frac{2}{3}$ on the number line. This step is done for you.

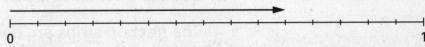

C. Draw an arrow to represent $\frac{1}{5}$ on the number line.

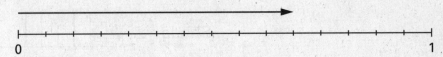

D. Count the number of sections between the end of the arrow for $\frac{2}{3}$ and the end of the arrow for $\frac{1}{5}$.

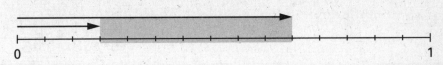

Since each section represents $\frac{1}{15}$, the difference is $\frac{\square}{15}$.

E. Complete. $\frac{2}{3} - \frac{1}{5} = \frac{\square}{15} - \frac{\square}{15}$

$$= \frac{\square}{15}$$

Reflecting

▶ Which do you find more useful to subtract fractions: fraction strips or number lines? Why?

_____

_____

Practising

Text page 318

5. a) Shade and label the fraction strips to explain how you know $\frac{4}{5} - \frac{1}{3}$ equals about $\frac{1}{2}$.

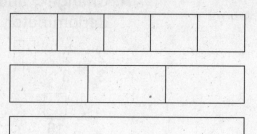

> **Hint**
>
> On the last fraction strip, shade the difference between the shaded areas on the first and second fraction strips.

b) Use the fraction strips below to subtract $\frac{4}{5} - \frac{1}{3}$.

The fraction strips are divided into 15 sections because a common denominator for $\frac{4}{5}$ and $\frac{1}{3}$ is 15.

Shade $\frac{4}{5}$. $\frac{4}{5} = \frac{\square}{15}$

Shade $\frac{1}{3}$. $\frac{1}{3} = \frac{\square}{15}$

Shade the difference.

Complete. $\frac{4}{5} - \frac{1}{3} = \frac{\square}{15} - \frac{\square}{15}$

$= \frac{\square}{15}$

6. Use the number line below to subtract $\frac{4}{3} - \frac{1}{2}$.

> **Hint**
>
> Remember, count the number of sections between the end of the arrow for $\frac{4}{3}$ and the end of the arrow for $\frac{1}{2}$.

Write a common denominator for $\frac{4}{3}$ and $\frac{1}{2}$. _____

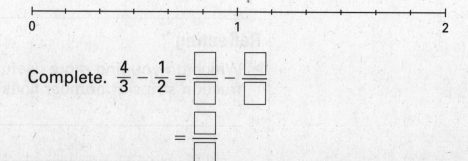

Complete. $\frac{4}{3} - \frac{1}{2} = \frac{\square}{\square} - \frac{\square}{\square}$

$= \frac{\square}{\square}$

8. Write a common denominator for $\frac{7}{10}$ and $\frac{1}{4}$. _____

Change $\frac{7}{10}$ and $\frac{1}{4}$ so they each have the common denominator.

$$\frac{7}{10} = \frac{7 \times \square}{10 \times \square} \qquad\qquad \frac{1}{4} = \frac{1 \times \square}{4 \times \square}$$

$$= \frac{\square}{\square} \qquad\qquad\qquad\quad = \frac{\square}{\square}$$

Shade the fraction strips to calculate $\frac{7}{10} - \frac{1}{4}$.

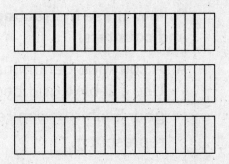

Complete. $\dfrac{7}{10} - \dfrac{1}{4} = \dfrac{\square}{\square} - \dfrac{\square}{\square}$

$$= \dfrac{\square}{\square}$$

9. Write a common denominator for $\frac{3}{5}$ and $\frac{1}{10}$. _____

Use the number line to calculate $\frac{3}{5} - \frac{1}{10}$.

Complete. $\dfrac{3}{5} - \dfrac{1}{10} = \dfrac{\square}{\square} - \dfrac{\square}{\square}$

$$= \dfrac{\square}{\square}$$

END

Subtracting Fractions with Grids

▶ **GOAL** Subtract fractions using grids and counters.

Problem 1

Follow these instructions to subtract $\frac{2}{3} - \frac{1}{4}$ using a grid and counters.

A. A common denominator for $\frac{2}{3}$ and $\frac{1}{4}$ is 12.
Use a 3-by-4 grid because there are 12 squares in the grid. Each row can represent $\frac{1}{3}$ and each column can represent $\frac{1}{4}$.

B. Place counters in the cells to represent $\frac{2}{3}$ of the grid. This step is done for you. ▶

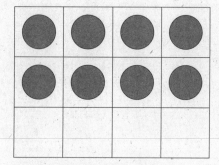

How many rows represent $\frac{2}{3}$ of the grid? _____
How many counters are on the grid? _____

C. Rearrange the counters to fill as many columns as possible. This step is done for you. ▶

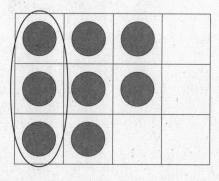

D. Remove the number of counters that represent $\frac{1}{4}$. The counters that should be removed are circled for you.

How many counters are left? _____
What fraction of the grid does this represent? $\frac{\Box}{\Box}$

E. Complete. $\frac{2}{3} - \frac{1}{4} = \frac{\Box}{12} - \frac{\Box}{12}$

$= \frac{\Box}{12}$

Problem 2

Follow these instructions to subtract $\frac{7}{8} - \frac{3}{4}$ using a grid and counters.

A. Place counters in the cells to show $\frac{7}{8}$.

How many counters did you use?

_____

B. How many rows represent $\frac{3}{4}$?

_____

C. Remove the number of counters that represent $\frac{3}{4}$.

How many counters are left?

_____

What fraction of the grid does this represent? $\frac{\square}{\square}$

D. Complete. $\frac{7}{8} - \frac{3}{4} = \frac{\square}{8} - \frac{\square}{8}$

$= \frac{\square}{8}$

Reflecting

▶ How do the grids used in Problems 1 and 2 show the common denominators of the fractions?

_____

_____

_____

_____

_____

TURN

Practising

Text page 324

5. Explain why a 3-by-5 grid models the common denominator of $\frac{2}{3}$ and $\frac{1}{5}$. Give two reasons.

_____

_____

_____

Use the grid and counters to model and calculate the difference.

Complete.

$$\frac{2}{3} - \frac{1}{5} = \frac{\boxed{}}{15} - \frac{\boxed{}}{15}$$

$$= \frac{\boxed{}}{\boxed{}}$$

6. Susan has $\frac{7}{12}$ of a movie left to watch. She watched $\frac{1}{3}$ of the movie on Sunday.

Use the grid and counters to show how much of the movie Susan still has to watch.

Complete. $\dfrac{7}{12} - \dfrac{1}{3} = \dfrac{\boxed{}}{12} - \dfrac{\boxed{}}{12}$

$= \dfrac{\boxed{}}{\boxed{}}$

7. What is a common denominator for $\frac{4}{5}$ and $\frac{2}{3}$? _____

What size grid will model $\frac{4}{5} - \frac{2}{3}$? Why?

_____

_____

Cut out a grid from 2 cm grid paper.
Tape it in the space below.
Use the grid and counters to model and calculate $\frac{4}{5} - \frac{2}{3}$. Complete the subtraction equation.

$\dfrac{4}{5} - \dfrac{2}{3} = \dfrac{\boxed{}}{\boxed{}} - \dfrac{\boxed{}}{\boxed{}}$

$= \dfrac{\boxed{}}{\boxed{}}$

Adding and Subtracting Mixed Numbers

You will need
- a ruler
- counters

▶ **GOAL** Add and subtract mixed numbers using models.

Method 1

Follow these instructions to add $2\frac{1}{2} + 1\frac{4}{5}$ using grids and counters.

A. $2\frac{1}{2} + 1\frac{4}{5}$ can be written as $2 + 1 + \frac{1}{2} + \frac{4}{5}$.

Add the fractions $\frac{1}{2}$ and $\frac{4}{5}$ first.

Use 2-by-5 grids to show the common denominator for $\frac{1}{2}$ and $\frac{4}{5}$.

Use counters to show $\frac{1}{2}$ on one grid and $\frac{4}{5}$ on the other grid. Each grid represents 1 whole.

$\frac{1}{2}$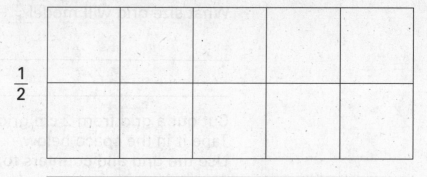

$\frac{4}{5}$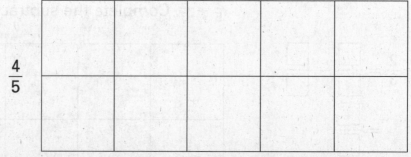

How many counters in total did you use? _____

Write the number of counters as a fraction. $\frac{\boxed{}}{10}$

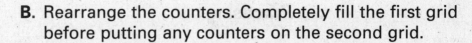

B. Rearrange the counters. Completely fill the first grid before putting any counters on the second grid.

Write the number of counters as a mixed number. $\square\dfrac{\square}{\square}$

C. Complete. $2\dfrac{1}{2} + 1\dfrac{4}{5} = 2 + 1 + \dfrac{\square}{10} + \dfrac{\square}{10}$

$$= 2 + 1 + \dfrac{\square}{10}$$

$$= 2 + 1 + \underline{\quad}\dfrac{\square}{10}$$

$$= \underline{\quad}\dfrac{\square}{10}$$

Method 2

Follow these instructions to subtract $3 - 1\dfrac{3}{4}$ using a number line.

A. Draw arrows on the number line to show 3 and $1\dfrac{3}{4}$. This step is done for you.

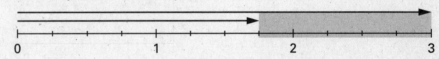

B. Count the number of sections between the end of the arrow for 3 and the end of the arrow for $1\dfrac{3}{4}$. _____

Write the number of spaces as a fraction. $\dfrac{\square}{4}$

C. Complete. Write the answers as an improper fraction and as a mixed number.

$$3 - 1\dfrac{3}{4} = \dfrac{\square}{4}$$

$$= \underline{\quad}\dfrac{\square}{\square}$$

Reflecting

▶ How is using a number line to model a difference of mixed numbers different from using it to model a difference of proper fractions?

_____

TURN

Practising

Text page 328 **7.** Use grids and counters to add $1\frac{1}{2} + 1\frac{7}{8}$.

Write a common denominator for $\frac{1}{2}$ and $\frac{7}{8}$. _____

Draw grids to show the common denominator.
Then use counters to add the fractions.

$\frac{1}{2}$

$\frac{7}{8}$

Complete.

$$1\frac{1}{2} + 1\frac{7}{8} = 1 + 1 + \frac{\Box}{8} + \frac{7}{8}$$

$$= 1 + 1 + \frac{\Box}{8}$$

$$= 1 + 1 + \underline{}\frac{\Box}{8}$$

$$= \underline{}\frac{\Box}{8}$$

Hint

Remember, after you add the fractions, change the improper fraction to a mixed number.

9. Use the number lines to subtract. Show your work.

a) $6 - 1\frac{4}{5} = \dfrac{\boxed{}}{\boxed{}}$

$= \underline{}\dfrac{\boxed{}}{\boxed{}}$

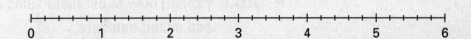

Explain why there are 5 sections between each whole number.

_____

b) $3 - 1\frac{9}{10} = \dfrac{\boxed{}}{\boxed{}}$

$= \underline{}\dfrac{\boxed{}}{\boxed{}}$

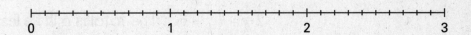

c) $4 - \frac{7}{8} = \dfrac{\boxed{}}{\boxed{}}$

$= \underline{}\dfrac{\boxed{}}{\boxed{}}$

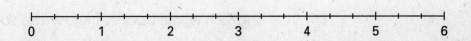

d) $6 - 4\frac{2}{3} = \dfrac{\boxed{}}{\boxed{}}$

$= \underline{}\dfrac{\boxed{}}{\boxed{}}$

END

Communicating about Estimation Strategies

▶ **GOAL** Explain how to estimate sums and differences of fractions and mixed numbers.

Problem

Simon needs $2\frac{1}{3}$ boards for part of a deck and $4\frac{1}{2}$ for another part of the deck. About how many boards does Simon need?

Simon's Explanation

I estimated that $2\frac{1}{3}$ is a little less than $2\frac{1}{2}$.

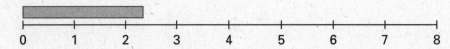

$2\frac{1}{2} + 4\frac{1}{2} = 6\frac{2}{2}$. The total is a little less than 7 boards.

Tien used the Communication Checklist to help Simon improve his explanation.

Tien's Suggestions

• Show both boards on the number line.

• Explain why the total is a little less than 7 boards.

Simon's New Explanation

I estimated that $2\frac{1}{3}$ is a little less than $2\frac{1}{2}$.

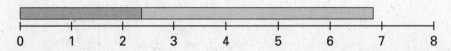

I know that $2\frac{1}{2} + 4\frac{1}{2}$ is $6\frac{2}{2}$. That is 7 whole boards, since $\frac{2}{2}$ is another whole board. I know the total is a little less than 7 boards, since $2\frac{1}{3}$ is a little less than $2\frac{1}{2}$.

Reflecting

▶ Why is a model helpful for explaining an estimation strategy to someone else?

_____

_____

_____

_____

Practising

Text page 332

5. George's family has $5\frac{1}{2}$ packages of crackers. On Sunday, George's cousins ate $1\frac{5}{6}$ packages of crackers. About how many packages are left?

a) Circle the operation you have to perform.

   addition   subtraction

b) Draw a picture to estimate the calculation.

c) Use the Communication Checklist on page 218 to help explain your estimation.

_____

_____

_____

_____

9.8
Text page 334

Adding and Subtracting Using Equivalent Fractions

▶ **GOAL** Develop a method for adding and subtracting fractions without using models.

MATH TERM

equivalent fractions fractions that have the same value but have different-sized sections

Use these instructions to find **equivalent fractions** for $\frac{2}{3}$.

A. Divide each third of the fraction strip above into two equal parts. This step is shown below.

How many parts are there now? _____

Write an equivalent fraction for $\frac{2}{3}$. $\dfrac{\square}{6}$

B. Complete the equivalent fraction equation.

$$\frac{2}{3} = \frac{2 \times \square}{3 \times \square}$$
$$= \frac{\square}{6}$$

Use what you know about changing thirds into sixths to solve this subtraction question.

$$\frac{2}{3} - \frac{1}{6} = \frac{2 \times \square}{3 \times \square} - \frac{1}{6}$$
$$= \frac{\square}{6} - \frac{1}{6}$$
$$= \frac{\square}{6}$$

Follow these instructions to use equivalent fractions to calculate $\frac{2}{3} + \frac{1}{2}$.

A. Write a common denominator for $\frac{2}{3}$ and $\frac{1}{2}$. _____

B. Break each third into two equal parts to make sixths.

Write an equivalent fraction for $\frac{2}{3}$. $\dfrac{\boxed{}}{6}$

C. Break each half into three equal parts to make sixths.

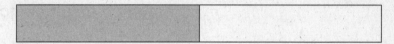

Write an equivalent fraction for $\frac{1}{2}$. $\dfrac{\boxed{}}{6}$

D. Complete the addition.

$$\frac{2}{3} + \frac{1}{2} = \frac{2 \times \boxed{}}{3 \times \boxed{}} + \frac{1 \times \boxed{}}{2 \times \boxed{}}$$

$$= \frac{\boxed{}}{6} + \frac{\boxed{}}{6}$$

$$= \frac{\boxed{}}{6}$$

$$= \underline{} \frac{\boxed{}}{6}$$

Hint

Write the improper fraction as a mixed number.

Reflecting

► Why are equivalent fractions useful for adding and subtracting fractions?

_____

_____

_____

_____

TURN

Practising

Text page 336

5. What common denominator can you use to add or subtract each pair of fractions?

a) $\frac{\blacksquare}{4}$ and $\frac{\blacksquare}{6}$

common denominator: _____

b) $\frac{\blacksquare}{8}$ and $\frac{\blacksquare}{16}$

common denominator: _____

c) $\frac{\blacksquare}{5}$ and $\frac{\blacksquare}{7}$

common denominator: _____

d) $\frac{\blacksquare}{4}$ and $\frac{\blacksquare}{9}$

common denominator: _____

6. Add. Show your work.

a) $\frac{5}{8} + \frac{1}{4} = \frac{5}{8} + \frac{1 \times \square}{4 \times \square}$

$= \frac{5}{8} + \frac{\square}{8}$

$= \frac{\square}{8}$

b) $\frac{3}{4} + \frac{7}{10} = \frac{3 \times \square}{4 \times \square} + \frac{7 \times \square}{10 \times \square}$

$= \frac{\square}{\square} + \frac{\square}{\square}$

$= \frac{\square}{\square}$

$= \underline{\quad} \frac{\square}{\square}$

7. Subtract. Show your work.

a) $\dfrac{5}{8} - \dfrac{1}{4} = \dfrac{5}{8} - \dfrac{1 \times \boxed{}}{4 \times \boxed{}}$

$= \dfrac{\boxed{}}{8} - \dfrac{\boxed{}}{8}$

$= \dfrac{\boxed{}}{\boxed{}}$

b) $\dfrac{3}{4} - \dfrac{7}{10} = \dfrac{3 \times \boxed{}}{4 \times \boxed{}} - \dfrac{7 \times \boxed{}}{10 \times \boxed{}}$

$= \dfrac{\boxed{}}{\boxed{}} - \dfrac{\boxed{}}{\boxed{}}$

$= \dfrac{\boxed{}}{\boxed{}}$

10. Add or subtract. Show your work.

a) $\dfrac{1}{9} + \dfrac{1}{3} = \dfrac{1}{9} + \dfrac{1 \times \boxed{}}{3 \times \boxed{}}$

$= \dfrac{\boxed{}}{\boxed{}} + \dfrac{\boxed{}}{\boxed{}}$

$= \dfrac{\boxed{}}{\boxed{}}$

b) $\dfrac{1}{2} - \dfrac{1}{4} = \dfrac{1 \times \boxed{}}{2 \times \boxed{}} - \dfrac{1}{4}$

$= \dfrac{\boxed{}}{\boxed{}} - \dfrac{\boxed{}}{\boxed{}}$

$= \dfrac{\boxed{}}{\boxed{}}$

END

10.1 Building and Packing Prisms

Text page 348

You will need
- nets of prisms
- scissors
- tape

- a protractor

▶ **GOAL** Build prisms from nets.

Imagine your job is to design efficient packaging. How can you design packages that will pack into cartons with no gaps between them?

Step 1: With your group, cut out six nets for a regular triangle-based prism. Fold and tape the nets to construct six prisms.

Step 2: Arrange the prisms so the bases have one common **vertex**, like in the diagram below.

MATH TERM

vertex
the point at the corner of an angle or shape

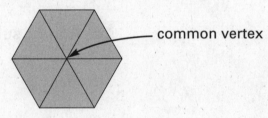

common vertex

Do the prisms pack with no gaps? _____

Step 3: Count the number of angles at the vertex where the prisms meet.

Number of angles: _____

Step 4: Use a protractor to measure each angle at the common vertex.

Measure of each angle: _____

Step 5: Add all the angle measures.

Sum of angle measures: _____

Hint

You will need:
- four square-based prism nets
- three pentagon-based prism nets
- three hexagon-based prism nets

Step 6: Repeat Steps 1 to 5 for each of the following prisms:
- square-based prism
- regular pentagon-based prism
- regular hexagon-based prism

Fill in the information for each prism in this table.

Prism base	Do the prisms pack with no gaps?	Number of angles at vertex where prisms meet	Measure of angles	Sum of angles at vertex where prisms meet
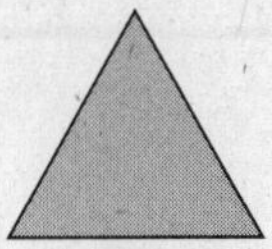 △				
▢				
 ⬠				
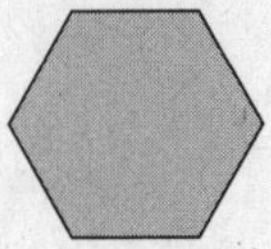 ⬡				

What is the same about the types of prisms that pack with no gaps?

_____

_____

_____

Reflecting

▶ How is packing a prism similar to tiling a plane?

_____

_____

_____

▶ How is packing a prism different from tiling a plane?

_____

_____

Building Objects from Nets

You will need
- Cutout Page 10.2
- scissors
- tape

▶ **GOAL** Build 3-D shapes from nets.

Follow these instructions to explore nets that make a cube.

A. Look at the picture of the cube in the margin. Add the following labels to the cube's net below: front, left side, top, and back.

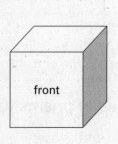

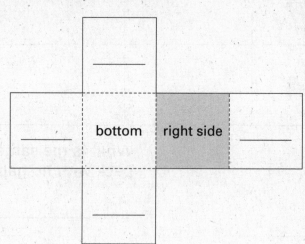

B. Circle the nets below that would form a cube.

1   **2**

Hint
Cut out these nets on Cutout Page 10.2 and see if you can make a cube from each net.

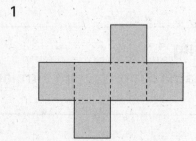

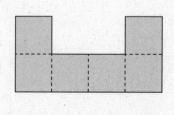

3   **4**

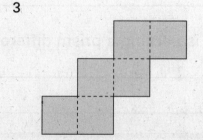

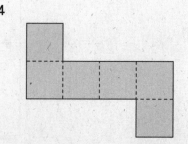

C. Explain why you cannot make a cube from the net(s) you did not circle in Step 2.

_____

_____

_____

D. Look for the similar pattern in nets 1 and 4 in Step 2. Use the grid below to draw and shade two more nets that fit this pattern.

Reflecting

▶ How many faces will the net of a cube always have? _____

▶ How would the nets in Part D change if the cube was open at the top?

_____

_____

▶ Without constructing it, explain why it is impossible to construct a cube from the net below.

_____

_____

_____

TURN

Practising

Text page 352

5. Look for the geometric shapes in the model of the house.

polyhedron
a 3-D shape with faces that all have straight sides, such as a square-based pyramid or a triangle-based prism

a) What **polyhedrons** do you see?

_____

b) How many edges does the base of the model have?

_____

c) How many faces does the model have? _____

Hint

First, decide which face must represent the base of the house.

6. Circle the net that will fold to make the model in question 5.

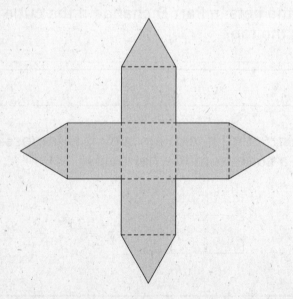

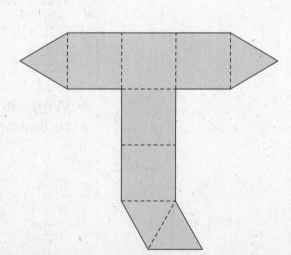

7. (Circle) the net(s) that will fold to make this rectangular prism. ▶
In each diagram, a possible base of the prism has been shaded.

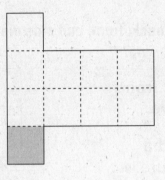   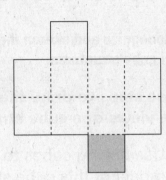   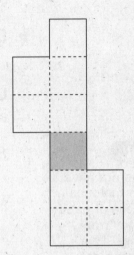

8. a) (Circle) the net below that will fold to make this triangle-based pyramid. ▶
Use your visualization skills to decide.

Hint

Cut out these nets on Cutout Page 10.2. Fold each net to check your decision.

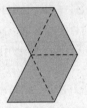   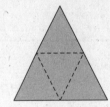

b) (Circle) the net below that will fold to make this shape. ▶
Use your visualization skills to decide.

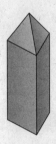

Hint

Cut out these nets on Cutout Page 10.2. Fold each net to check your decision.

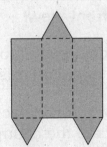   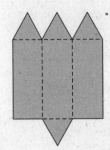

10.3
Text page 354

Top, Front, and Side Views of Cube Structures

You will need
- a building mat
- linking cubes

▶ **GOAL** Recognize and sketch the top, back, front, and side views of cube structures.

Use these steps to draw the top, front, and side views of a cube structure.

Step 1: Use linking cubes to build a model of this cube structure. ▶

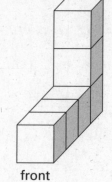

front

Hint

If you do not have a building mat, use a plain piece of paper and write "front," "left side," "back," and "right side" along each side of the page, as below.

```
        back
left side    right side
        front
```

Step 2: Place your model on the building mat. Look directly down on the model to see the top view.

Shade squares to draw the top view on this grid. ▶
This step is done for you.

Step 3: Draw a thick black line to indicate the change in depth in the top view.
This step is done for you.

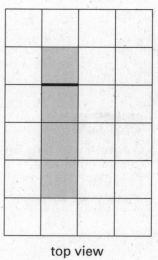

top view

Step 4: Bring your eyes level with the front of the model so that you are looking directly at the front view.

Shade squares to draw the front view on this grid. ▶

Hint

Each square represents a cube.

Hint

To identify a change in depth, count the number of cubes behind each square in your drawing.

Step 5: Draw a thick black line to indicate the change in depth in the front view.

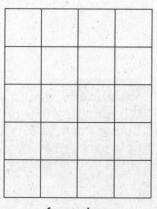

front view

230 Lesson 10.3: Top, Front, and Side Views of Cube Structures

Copyright © 2006 Nelson

Step 6: Bring your eyes level with the right side of the model so that you are looking directly at the right-side view.

Shade squares to draw the right-side view on this grid. ▶

Is there a change in depth in this view?

_____

right-side view

Reflecting

▶ What relationship do you see between the width of the top view and the width of the front view? Use the drawings from Steps 2 and 4.

_____

▶ What relationship do you see between the height of the front view and the height of the right-side view?

_____

▶ What does the right-side view tell you about the change in depth shown on the front view?

_____

_____

_____

_____

_____

_____

Practising

Text page 356

5. a) Build this structure using linking cubes. ▶

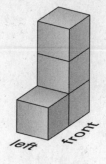

b) Place your structure on the building mat. Visualize what the structure would look like from the top, front, and left side.

c) Label each of the following diagrams as top, front, and left-side view.

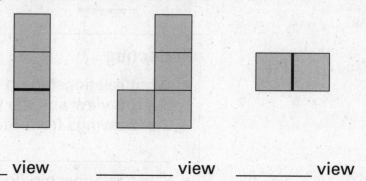

_____ view _____ view _____ view

d) Rotate your building mat to look at the right side of the structure.

Draw the right-side view on this grid.

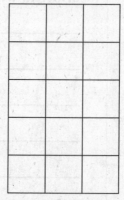

right-side view

e) Circle the view(s) that show a change in depth.

top view   front view   right-side view   left-side view

6. a) Build this structure with linking cubes. ▶

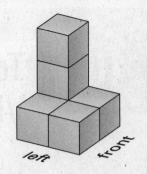

b) Look straight down on your structure to see the top view. Draw the top view on the first grid below.

Hint

Remember to draw a thick black line on any view where there is a change in depth.

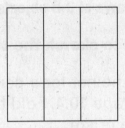

top view

front view

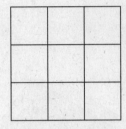

left-side view

c) Rotate your structure to see the front view. Draw the front view on the second grid above.

d) Rotate your structure to see the left-side view. Draw the left-side view on the third grid above.

Text page 357

8. a) Use these top, front, and side views below to build a linking-cube structure.

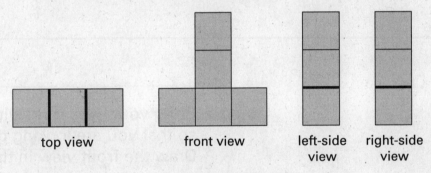

top view   front view   left-side view   right-side view

b) Draw the back view on this grid.

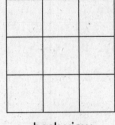

back view

10.4

Top, Front, and Side Views of 3-D Objects

You will need
- Cutout Page 10.4
- a building mat
- scissors
- tape
- a ruler

▶ **GOAL** Recognize and sketch the top, front, and side views of 3-D objects.

Use these steps to sketch the views of a 3-D object.

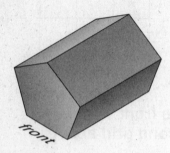

Step 1: Cut out the net for a pentagonal prism from Cutout Page 10.4. Fold the net and use tape to build the prism.

Step 2: Place your model on the building mat. Look directly down on the model to see the top view. Draw the top view in the box below.
Show any change in depth with a thick black line.

top view

Step 3: Bring your eyes level with the front of the model so that you are looking directly at the front view. Draw the front view in the box below.

front view | back view

Hint

If you need help drawing the front or back view, trace the face of your model.

Step 4: Bring your eyes level with the back of the model so that you are looking directly at the back view. Draw the back view in the box above.

Step 5: Bring your eyes level with the right side of the model so that you are looking directly at the right-side view.
Draw the right-side view in the box below.

right-side view

Hint
Remember to include a thick black line to show a change in depth.

How would the left-side view compare to the right-side view?

____

Reflecting

- How does the back view compare to the front view?

____

- Could a different face of your model have been the front view? How do you know?

____

____

- Why is it important to keep the object you are viewing in the same position on the building mat to draw different views?

____

____

____

TURN

Text page 362   **4.** Three views of a polyhedron are shown below.

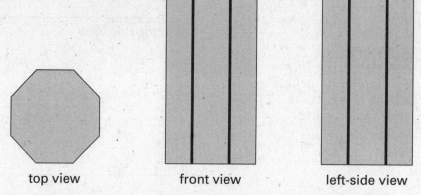

top view   front view   left-side view

a) How does the top view help to explain the thick black lines in the other two views?

_____

_____

_____

b) Name the polyhedron that has these views.

_____

5. Draw the views of this shape ▶ in the boxes below. Remember to add thick black lines when necessary.

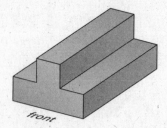

front

top view	front view	left-side view

6. Three views of a polyhedron are shown below.

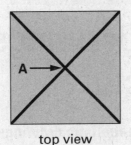 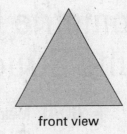 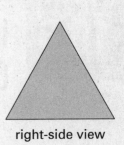

top view   front view   right-side view

a) What do the thick black lines on the top view tell you about the shape?

_____

_____

b) What do the front and right views tell you about point A marked on the top view?

_____

_____

c) Name the polyhedron that has these views.

_____

8. Three views of a polyhedron are shown below.

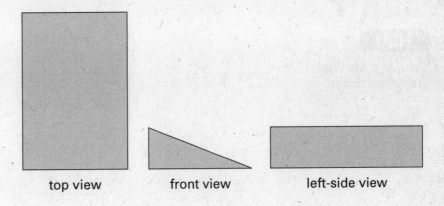

top view   front view   left-side view

Hint

Look at the shapes of the views for clues.

Name the polyhedron that has these views.

_____

10.5 Isometric Drawings of Cube Structures

Text page 364

You will need
- linking cubes

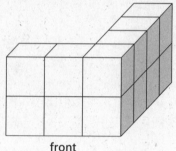

- a ruler

▶ **GOAL** Make realistic drawings of cube structures on triangle dot paper.

Follow these instructions to make an isometric drawing of a cube structure.

A. Build a model of this cube structure using linking cubes. ▶

MATH TERM

isometric drawing
a drawing that represents a 3-D view of an object; it is drawn on triangle dot paper

B. An **isometric drawing** of the cube structure has been started below. Using your model, explain the lengths of sides A, B, and C. Side A is done for you.

Side A: There are three cubes along the front.

Side B: _____

Side C: _____

Hint

Vertical edges are drawn vertically.

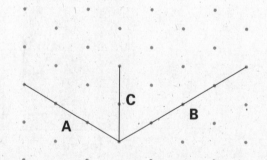

C. Complete the bottom layer of the cube structure on the grid. Draw lines for each cube.

D. Complete the drawing by adding the second layer of the cube structure. Erase any lines that you cannot see when looking at the cube structure.

238 Lesson 10.5: Isometric Drawings of Cube Structures

Copyright © 2006 Nelson

Using linking cubes, build a model of the cube structure in the isometric drawing below.

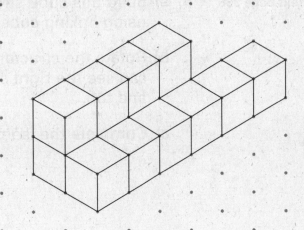

Could someone else build a different model of the cube structure and still be correct? Explain.

_____

_____

_____

Reflecting

▶ Explain why an isometric drawing does not always show all the details of a model.

_____

_____

_____

_____

TURN

Practising

Text page 366

4. a) Build this cube structure using linking cubes. ▶

b) Rotate the structure so you can see the right side, front, and top.

c) Complete the isometric drawing below.

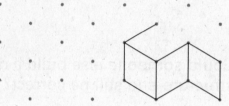

5. a) Build this cube structure using linking cubes. ▶

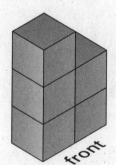

front

b) Rotate the structure so you can see the left side and the front.

c) In the first box, draw the bottom layer of the structure.

d) In the second box, draw the bottom two layers of the structure.

e) In the third box, draw all three layers to complete the isometric drawing.

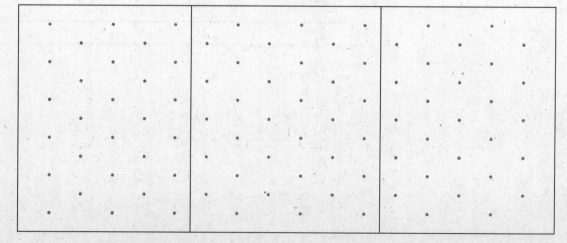

7. Build the cube structures shown in the margin.
Then make an isometric drawing of each structure.

a)

b)

c)

END

Isometric Drawings of 3-D Objects

You will need
- a book
- a 3-D model of a triangle-based prism

- a ruler

▶ **GOAL** Make realistic drawings of 3-D objects on triangle dot paper.

Use these steps to make an isometric drawing of a book.

Step 1: Rotate your book so that one corner is pointing at you.

Step 2: Draw the two base edges as indicated on the picture in the margin. Estimate the measurements.

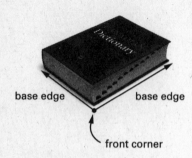

base edge   base edge

front corner

Hint

Remember, vertical edges are drawn vertically.

Step 3: Draw the front face of the prism.

Step 4: Add the left-side face.

Step 5: Add the top face to complete the drawing.

front corner of book

Reflecting

▶ How is making an isometric drawing of a 3-D object different from making an isometric drawing of a cube structure?

_____

_____

_____

_____

Practising

Text page 370

Hint
Use a 3-D model of
the prism for help.

5. Make an isometric drawing of the triangle-based
prism by following the steps below. Estimate the
measurements.

Step 1: Choose the triangular base
as the front face of the
prism.
Step 2: Draw the base lines from
the front corner. This step
is done for you.
Step 3: Draw the front triangular
face of the prism.
Step 4: Complete the remaining
visible face(s).

6. Make an isometric drawing of
this triangle-based prism. ▶
Use the measurements given
in the diagram. Follow the
steps you used in question 5.

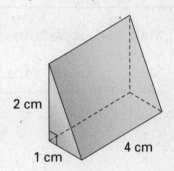

2 cm

1 cm

4 cm

END

Communicating about Views

You will need
- linking cubes

- a ruler

▶ **GOAL** Use mathematical language to describe views of 3-D objects.

Ryan wrote the following steps for making an isometric drawing of a rectangular prism.

Complete each of Ryan's steps on the triangle dot paper.

Step 1: Choose the rectangular base as the front face of the prism.

Step 2: Draw the base edges from the front corner.
One base edge should be three spaces.
The other base edge should be five spaces.

Step 3: Complete the front face of the prism.

Step 4: Complete the side face.

Step 5: Complete the top face.

Reflecting

▶ What information were you missing when you tried to draw the prism?

_____

_____

_____

_____

_____

Practising

Text page 374   7.

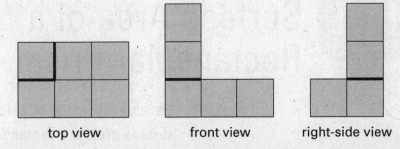

top view   front view   right-side view

a) Write steps in the box below to explain how to use these top, front, and right-side views to build a linking-cube structure.

b) What is good about your writing? Use the Communication Checklist to help you decide.

_____

_____

_____

c) What can you improve in your steps? Use the Communication Checklist to help you decide.

_____

_____

Communication Checklist

❑ Did you explain your thinking?

❑ Did you give all the important details?

❑ Did you use sketches to make your thinking clear?

❑ Did you use proper math language?

d) Check your steps by building the structure with linking cubes. Make any changes you think would make the instructions clearer.

Lesson 10.7: Communicating about Views **245**

END

Surface Area of a Rectangular Prism

▶ **GOAL** Develop a formula to calculate the surface area of a rectangular prism.

Examine a rectangular prism.
How many surfaces does it have? _____

Name the surfaces. Two surfaces are named for you.

front, back, _____

Follow these instructions to develop a formula to calculate the surface area of a rectangular prism.

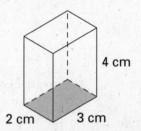

4 cm

2 cm   3 cm

A. Draw a sketch of each of this rectangular prism's surfaces on the grid. The bottom is done for you.

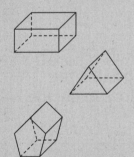

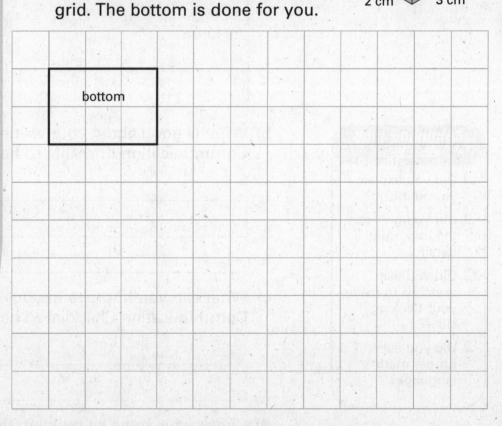

bottom

B. Write the area of each surface on the grid.

C. Calculate the total area of all the surfaces.

_____ + _____ + _____ + _____ + _____ + _____ = _____

So, the **surface area** of the rectangular prism is
_____ cm^2.

D. Name the surfaces of the rectangular prism that are
congruent (identical). Show how to calculate their areas.

_____ and _____ (Area = ____ × ____ = ____ cm^2)

_____ and _____ (Area = ____ × ____ = ____ cm^2)

_____ and _____ (Area = ____ × ____ = ____ cm^2)

E. You need to calculate the area of each pair of surfaces
(from Part D) only once and then multiply each area
by 2. Add these products together to calculate the
surface area of a rectangular prism.
Fill in the blanks to show how to do this.

Surface Area

= 2 ×(____ × ____) + 2 ×(____ × ____) + 2 ×(____ × ____)

= 2 ×(____) + 2 ×(____) + 2 ×(____)

= ____ + ____ + ____

= ____ cm^2

Reflecting

▶ Explain why a formula for calculating the surface
area of a rectangular prism is

Surface Area = 2 × (sum of areas of the 3 different
surfaces)

_____

_____

_____

_____

Practising

Text page 387

6. Write the length, width, and height of each prism. Then calculate each surface area.

a) length = _____ cm
width = _____ cm
height = _____ cm

Surface Area =

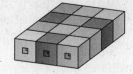

b) length = _____ cm
width = _____ cm
height = _____ cm

Surface Area =

c) length = _____ cm
width = _____ cm
height = _____ cm

Surface Area =

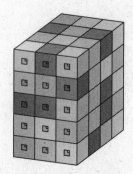

8. Calculate the surface area of this cube.

Surface Area =

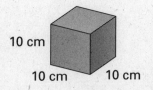

10 cm

10 cm   10 cm

9. **a)** Use 12 centimetre linking cubes to build as many different rectangular prisms as possible. List their dimensions and surface areas in the table.

Length (cm)	Width (cm)	Height (cm)	Surface Area (cm²)

b) List the dimensions of the rectangular prism with the greatest surface area.

length = _____ cm
width = _____ cm
height = _____ cm

c) List the dimensions of the rectangular prism with the least surface area.

length = _____ cm
width = _____ cm
height = _____ cm

d) How do the dimensions of the least surface area compare to the dimensions of the greatest surface area?

_____

_____

END

Volume of a Rectangular Prism

You will need
- centimetre linking cubes
- a calculator
- a ruler

▶ **GOAL** Develop a formula to calculate the volume of a rectangular prism.

James is packing small dice into a box like this.▶
Each edge of each die is 1 cm.
How many dice will the box hold?

Follow these instructions to develop a formula to calculate the volume of the box.

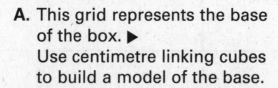

height = 4 cm
length = 5 cm
width = 3 cm

A. This grid represents the base of the box. ▶
Use centimetre linking cubes to build a model of the base.

How many linking cubes are in the base? _____

Area of base = _____ cm²

5 cm

3 cm

B. Stack cubes to build a model of the box.
How many layers did you need? _____
How many linking cubes are in each layer? _____
How many linking cubes are in the box in total?

_____ × _____ = _____

Volume of box = _____ cm³

C. Explain why this is a formula for the volume of a rectangular prism (a box):

Volume = Area of base × height

_____

_____

_____

D. Turn your model so the base matches the base on this grid. ▶

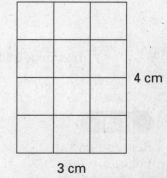

4 cm

3 cm

Area of base

= _____ cm × _____ cm

= _____ cm²

Height = _____ cm

Volume

= Area of base × height

= _____ cm² × _____ cm

= _____ cm³

E. Turn your model once more so the base matches the base on this grid. ▶

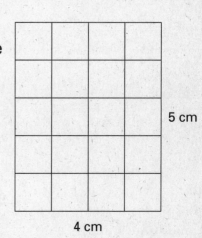

5 cm

4 cm

Area of base

= _____ cm × _____ cm

= _____ cm²

Height = _____ cm

Volume

= Area of base × height

= _____ cm² × _____ cm

= _____ cm³

Look at the diagram on page 250 to find the height when the length is 3 cm and the width is 4 cm.

Hint

Reflecting

▶ Compare the three volume calculations. How are they the same and how are they different?

_____

_____

_____

▶ Write a different formula to calculate the volume of a rectangular prism. Use the words length, width, and height.

Volume = _____

TURN ➡

Practising

Text page 390 **7.** Calculate the volume of each prism.

a)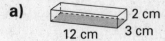
2 cm
3 cm
12 cm

> **Hint**
>
> Remember,
> Volume = length ×
> width × height.

Volume = _____ cm × _____ cm × _____ cm

= _____ cm³

b)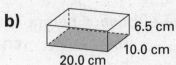
6.5 cm
10.0 cm
20.0 cm

Volume = _____ cm × _____ cm × _____ cm

= _____ cm³

c)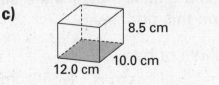
8.5 cm
10.0 cm
12.0 cm

Volume = _____ cm × _____ cm × _____ cm

= _____ cm³

9. a) Sketch a rectangular prism with these dimensions:

length = 4 cm, width = 2 cm, height = 3 cm

> **Hint**
>
> See Lesson 10.5 on
> page 238 to refresh
> your memory about
> making isometric
> drawings.

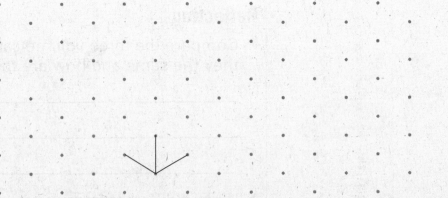

b) Calculate the volume of the prism.

Volume = _____ cm × _____ cm × _____ cm

= _____ cm³

Connect Your Work

A rectangular prism has a volume of 24 cm³.

Follow these instructions to find three sets of dimensions that will result in this volume.

A. Start with a base that has a length of 6 cm and a width of 2 cm. Build this base with linking cubes.

B. Calculate the area of the base.

Area of base = _____ cm × _____ cm

= _____ cm²

What height will give a volume of 24 cm³? _____ cm

Complete the drawing.

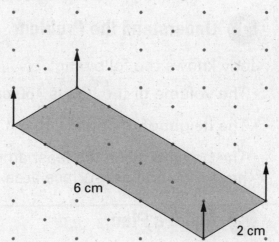

Dimensions
length = _____ cm
width = _____ cm
height = _____ cm

C. Build two other rectangular prisms with the 24 centimetre linking cubes.
Use all the cubes for each rectangular prism.
Record the dimensions of each rectangular prism in the table.

Length (cm)	Width (cm)	Area of base (cm²)	Height (cm)	Volume (cm³)
6	2			

END

Solve Problems by Guessing and Testing

You will need
• a calculator

▶ **GOAL** Guess and test and use a table to solve measurement problems.

Jody is designing a snack box. The box must be 10 cm high and have a volume of 450 cm³. She wants to use the least amount of cardboard possible.

How can she determine the best dimensions the box should have to waste the least amount of cardboard?

1 Understand the Problem

Jody knows the following:

• The volume of the box is 450 cm³.

• The height of the box is 10 cm.

• The box that uses the least amount of cardboard will have the smallest surface area.

2 Make a Plan

Jody knows the formula for the volume of a prism is

Volume = Area of base × height

Fill in the volume and height:

_____ cm³ = Area of base × _____ cm

What number × 10 = 450? _____

So, Area of base must be _____ cm²,
since 450 = _____ × 10.

Jody decides to use a calculator to find numbers for length and width that produce an area of 45 cm².

3 Carry Out the Plan

Jody organizes the lengths and widths in a table. She calculates the surface area for each set of dimensions.

Length (cm)	Width (cm)	Height (cm)	Surface Area (cm²) = 2 × (sum of areas of 3 different surfaces)
1.0	45.0	10.0	= 2 × (45.0 + 450 +10) = 2 × (505) = 1010.0
2.0	22.5	10.0	580.0
3.0	15.0	10.0	450.0
4.0	11.25	10.0	395.0
5.0	9.0	10.0	370.0
6.0	7.5	10.0	360.0
7.0	6.43	10.0	358.6
8.0	5.625	10.0	362.5

Hint

Jody uses a calculator and rounds her results.

Circle the surface area in the table that is the smallest.

4 Look Back

Do the surface area numbers that come *before* the circled one increase or decrease? _____

Do the surface area numbers that come *after* the circled one increase or decrease? _____

What do you notice about the length and width as you move toward the smallest surface area?

_____

Reflecting

▶ How can you use your observation about the length and width in Part 4 to improve your solution?

_____

_____

Practising

Text page 396

5. a) A box has a volume of 1500 cm³. Determine a possible length, width, and height for the box.

Hint

Use a calculator to find any three numbers that produce a product of 1500.

length = _____ cm
width = _____ cm
height = _____ cm

Show that these dimensions give a volume of 1500 cm³.

Volume = length × width × height
1500 cm³ = _____ cm × _____ cm × _____ cm
= _____ cm³

Calculate the surface area of this box.

Surface Area = 2 × (sum of areas of 3 different surfaces)
=
=
=
=

b) Determine a second set of dimensions for the box with a volume of 1500 cm³.

length = _____ cm
width = _____ cm
height = _____ cm

Calculate the surface area of this box.

Surface Area = 2 × (sum of areas of 3 different surfaces)
=
=
=
=

c) Which set of dimensions would be better if you wanted to use less cardboard to make a box with a volume of 1500 cm³? Explain.

_____

_____

7. Someone spilled juice on these plans for a playground. The length and width are missing from the plans.

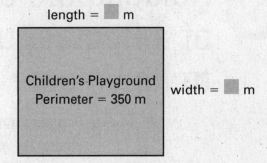

length = ☐ m

Children's Playground
Perimeter = 350 m

width = ☐ m

a) Determine a possible length and width that the playground might be.

length = _____ m
width = _____ m

Show that the length and width you chose result in a perimeter of 350 m.

Perimeter = 2 × (length + width)
350 m = 2 × (_____ m + _____ m)
= 2 × (_____ m)
= _____ m

Calculate the area of the playground.

Area = length × width
= _____ m × _____ m
= _____ m²

b) Determine a second set of length and width measurements that the playground might be.

length = _____ m
width = _____ m

Calculate the area of the playground.

Area = length × width
= _____ m × _____ m
= _____ m²

c) Which set of length and width measurements results in a smaller area for the playground?

11.4 Relating the Dimensions of a Rectangular Prism to Its Volume

You will need
- centimetre linking cubes
- a calculator
- a ruler

▶ **GOAL** Discover how changing the sides of a rectangular prism affects its volume.

Use these steps to explore how changing the sides of a rectangular prism affects its volume.

Step 1: Use 16 centimetre linking cubes to build a rectangular prism one layer high.
Record the volume in the table.

Model	Length (cm)	Width (cm)	Height (cm)	Volume (cm³)
	4	4	1	
	4	4	2	

Step 2: Add a second layer. Record the information for the new prism in the table.

Step 3: Continue adding layers. Record the information for each prism in the table. How does the volume change as the height increases?

Hint

Use the relationship between the volume and the height to calculate the volume.

_____

What is the volume when the height is 6 cm?
_____ cm³

Reflecting

▶ (Circle) two volumes in the table where one is double the other. Compare the heights. What do you notice?

_____

▶ Shade two volumes in the table where one is triple the other. Compare the heights. What do you notice?

_____

Connect Your Work

Follow these instructions to see if the volume of a rectangular prism doubles when you double *one* of the dimensions.

A. Build a 1 cm × 2 cm × 3 cm rectangular prism with centimetre linking cubes.

B. Double the width. Record the information for the new prism in the table.

C. Repeat Part B twice.

Length (cm)	Width (cm)	Height (cm)	Volume (cm³)
1	2	3	6
1	4	3	
1		3	
1		3	

Complete the conclusion:
When the width is doubled, the volume _____.

D. Repeat B and C, but this time, double the length.

Record the information for each prism in the table.

Length (cm)	Width (cm)	Height (cm)	Volume (cm³)
1	2	3	6
2	2	3	
	2	3	
	2	3	

Complete the conclusion:
When the height is doubled, the volume _____.

Practising

Text page 400

5. The volume of a rectangular prism is 12 cm³.
   The height of the prism is 2 cm.

   a) Determine a possible length and width for the
      prism.

      Volume = Area of base × height

      Fill in the volume and height:

      _____ cm³ = Area of base × _____ cm

      So, Area of base = _____ cm², since 12 = _____ × 2.

      Determine what two numbers multiplied by each
      other equal the area of the base.

      length = _____ cm
      width = _____ cm

   b) The height of the prism changes, causing the
      volume to become 24 cm³. The length and width
      remain the same.

      Determine the new height of the prism. _____ cm

      Draw a sketch of the rectangular prism with a
      volume of 12 cm³ and a rectangular prism with
      a volume of 24 cm³ to support your answer.
      Label the dimensions on the drawings.

6. A rectangular prism is 5 m high, 3 m long, and 6 m wide.

    a) Determine the volume of the prism.

        Volume = _____ m × _____ m × _____ m
               = _____ m³

    b) If the height of the prism doubles and the other dimensions stay the same, what happens to the volume?

        _____

        Calculate the new volume.

        Volume = 2 × _____ m³
               =

    c) The length of the original prism doubles and the height and width stay the same. Determine the volume of the prism now.

        Volume =

7. A box is 7.5 cm high, 10.2 cm long, and 2.0 cm wide.

    a) Determine the volume of the box.

        Volume =

    b) Determine the volume of the box when its height triples.

        Volume =

    c) Determine the volume of the box when its length is multiplied by 4.

        Volume =

Exploring the Surface Area and Volume of Prisms

> ▶ **GOAL** Investigate relationships between surface area and volume of cubes and other rectangular prisms.

MATH TERM

cube
a rectangular prism with six congruent square faces

Use these steps to try to find a rectangular prism with the same volume as a **cube** but with a smaller surface area.

Step 1: Complete the table for cubes. You can write your calculations on a scrap piece of paper.

Length (cm)	Width (cm)	Height (cm)	Surface Area (cm²) = 2 × (*l* × *w* + *l* × *h* + *w* × *h*)	Volume (cm³) = *l* × *w* × *h*
1	1	1	= 2 × (1 × 1 + 1 × 1 + 1 × 1) = 2 × (1 + 1 + 1) = 2 × 3 = 6	= 1 × 1 × 1 = 1
2	2	2		
3	3	3		
4	4	4		
5	5	5		
6	6	6		
7	7	7		
8	8	8		
9	9	9		
10	10	10		

Step 2: Circle the row in the table of the cube whose surface area and volume are the same. Write the dimensions below.

length = _____ cm
width = _____ cm
height = _____ cm

Step 3: Find the length, width, and height for four rectangular prisms whose volumes are also 216 cm³. Use 9 cm as the height.

Volume = Area of base × height

Fill in the Volume and height:

_____ cm³ = Area of base × _____ cm

So, Area of base = _____ cm²,
since 216 cm³ = _____ cm² × 9 cm.

Area = length × width.
So, _____ cm² = length × width.

Find four whole number length and width combinations that give this area.
Record the lengths and widths in the table.

Length (cm)	Width (cm)	Height (cm)	Surface Area (cm²)	Volume (cm³)
		9		216
		9		216
		9		216
		9		216

Step 4: Calculate the surface area of each rectangular prism you found. You can write your calculations on a scrap piece of paper. Record the surface areas in the table.

Reflecting

▶ How do the surface areas of these rectangular prisms compare to the surface area of the cube with volume 216 cm³?

_____

▶ Predict the dimensions of the shape with the least surface area for a volume of 27 cm³.

_____

Explain your prediction.

_____

_____

END

Exploring Probability

You will need
- a spinner
- a calculator

▶ **GOAL** Determine probability from an experiment.

On average, Omar gets one hit for every three times at bat. He played 20 games last year and was up to bat three times per game.

Use these steps to determine the probability that Omar will get two hits in a game.

Step 1: Using a spinner like the one in the margin, complete the table for 20 games. Spin the spinner three times to represent Omar's three at-bats per game.
Use **H** for hit and **N** for no hit. The first three games have been completed for you.

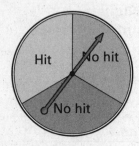

Game number	First at-bat	Second at-bat	Third at-bat
1	N	H	N
2	N	N	H
3	N	N	H
4			
5			
6			
7			
8			
9			
10			
11			
12			
13			
14			
15			
16			
17			
18			
19			
20			

All the possible outcomes for your experiment are listed in the table below. For example, in game 1, the outcome was no hit, hit, no hit. (You can write this as N-H-N).

| Outcome | Frequency | Experimental Probability | | | Class results |
		Fraction	Decimal	Percent	
N-H-N					
N-N-N					
N-N-H					
H-H-N					
N-H-H					
H-N-H					
H-N-N					
H-H-H					
Total	20	$\frac{20}{20}$	1.00	100%	100%

Step 2: Count the number of times each outcome occurred in the table on page 264. Enter the numbers in the Frequency column above.

MATH TERM

experimental probability
a measure of the likelihood of an event, based on data from an experiment

Step 3: Calculate the **experimental probability** using this ratio:

$$\frac{\text{frequency of an event}}{\text{total number of trials}}$$

Record the probability for each outcome as a fraction, decimal, and percent in the table above.

MATH TERM

favourable outcome
the result that you are investigating in a probability experiment

Step 4: Since you are investigating Omar getting two hits in a game, one **favourable outcome** would be H-H-N. Circle all favourable outcomes in the table above.

Step 5: Add the percent probabilities for the favourable outcomes to calculate the probability of Omar getting two hits in a game.

_____ + _____ + _____ = _____

Step 6: Complete the table on page 265 using the combined results of your class.

Calculate the probability of Omar getting two hits in a game based on the combined results of your class.

_____ + _____ + _____ = _____

Reflecting

▶ If you repeated the experiment, would your results be the same? Explain.

_____

_____

_____

▶ Why is it important to have a large number of trials when conducting a probability experiment?

_____

_____

_____

END

Calculating Probability

▶ **GOAL** Identify and state the theoretical probability of favourable outcomes.

When you toss a coin, there are two possible outcomes: Heads or Tails.

If the outcome you are interested in is Heads, you can calculate the **theoretical probability** of Heads being tossed using this ratio:

$$\frac{\text{favourable outcomes}}{\text{total possible outcomes}} = \frac{1}{2} \text{ (or 0.5, or 50\%)}$$

MATH TERM

theoretical probability
a measure of the likelihood of an event, based on calculations

You can write the probability of an event as P(event). The event is usually a number or a word. For example,

$$P(\text{Heads}) = \frac{1}{2}$$

A regular die shows the numbers from 1 to 6. How many possible outcomes are there when you roll the die? _____

What is the probability of rolling a 5? $P(5) = \dfrac{\Box}{\Box}$

What is the probability of rolling a 9? $P(9) = \dfrac{\Box}{\Box}$

Reflecting

▶ Suppose you toss a coin and the outcome is Heads. Does this affect the probability of Heads on the next toss? Explain.

_____

_____

_____

Practising

Text page 416

4. Write each of the following probabilities as a fraction.

    a) *P*(one Heads and one Tails with two coins) = $\dfrac{\square}{\square}$

    b) *P*(Queen of ♠) = $\dfrac{\square}{\square}$

5. There are 52 cards in a full deck. What is each probability?

    a) *P*(4) = $\dfrac{\square}{\square}$

    b) *P*(♥) = $\dfrac{\square}{\square}$

    c) *P*(any 4 or any ♥) = $\dfrac{\square}{\square}$

6. Suppose you toss three pennies.

a) List all the possible outcomes for the toss. The first two are done for you.

Coin 1	Coin 2	Coin 3	Outcome
H	T	H	H-T-H
H	H	T	H-H-T

b) What is the probability of all three pennies landing Heads?

$$P(\text{H-H-H}) = \dfrac{\square}{\square}$$

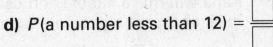

10. For the spinner below, what is each probability?

a) $P(\text{multiple of 3}) = \dfrac{\square}{\square}$

b) $P(\text{factor of 12}) = \dfrac{\square}{\square}$

c) $P(3, 5, \text{or } 8) = \dfrac{\square}{\square}$

d) $P(\text{a number less than 12}) = \dfrac{\square}{\square}$

END

Solve Problems Using Organized Lists

You will need
• a calculator

▶ **GOAL** Use organized lists to determine all possible outcomes.

Problem

Simon says to Rana, "I have exactly five coins in my hand worth a total of 50¢. There are no pennies. Guess what the coins are."

Use these steps to help Rana determine the probability of guessing correctly.

1 Understand the Problem

The following information is given in the problem:

• You have exactly five coins.

• The total value of the coins is 50¢.

• There are no pennies.

• There can be any combination of nickels, dimes, and quarters.

• There may be more than one of some types of coins.

• There may be none of some types of coins.

2 Make a Plan

Rana decides to make a table including all the possible combinations of coins that add up to 50¢ in an organized list.

Rana will make a table and use a system so she will not miss or repeat any combination.

3 **Carry Out the Plan**

Step 1: Complete Rana's table by listing all the combinations that add up to 50¢.

Quarters	Dimes	Nickels	Value	Total number of coins
2	0	0	50¢	2
1	2	1	50¢	4
1	1		50¢	
1	0		50¢	
0	5		50¢	
0	4		50¢	
0	3		50¢	
0	2		50¢	
0	1		50¢	
0	0		50¢	

Step 2: Circle the rows that have exactly five coins.

Step 3: What is the probability of Rana guessing the combination of coins if she has only one guess?

$$P(\text{correct guess}) = \frac{\boxed{}}{\boxed{}}$$

Hint

The rows you circled in Step 2 represent the possible correct guesses. There is only one correct guess.

4 **Look Back**

Rana sees a pattern in the numbers, so she is sure that she listed all the possible combinations.

Reflecting

▶ Describe the number patterns in Rana's list.

_____

_____

_____

_____

Practising

Text page 425

5. Suppose that you have two coins in your pocket. The coins can be pennies, nickels, dimes, or quarters.

a) Complete the table to determine how many different combinations are possible.
Use p to represent a penny, n to represent a nickel, d to represent a dime, and q to represent a quarter.

1st coin	2nd coin	Total value (¢)
p	p	
p	n	
p		
p		
n	n	
n	d	
n		
d		

Hint

Make sure you don't count the same combination twice. The position of the coin makes no difference; p, d is the same as d, p.

b) Describe the pattern in the table.

_____

_____

_____

_____

_____

c) How many of the combinations add up to less than 20¢? _____

d) What is the probability that the two coins in your pocket add up to less than 20¢?

$P(\text{value less than } 20¢) = \dfrac{\square}{\square}$

10. Samantha threw two darts and hit the dartboard each time. Both darts might have hit the same number.

a) Use an organized list to show all the possible scores Samantha might have.

1st dart	2nd dart	Total score
10	10	20
10		
10		
10		
5		

b) How many outcomes are there? _____

How many outcomes are less than 10? _____

What is the probability that her score will be less than 10?

P(less than 10) = $\dfrac{\Box}{\Box}$

Using Tree Diagrams to Calculate Probability

▶ **GOAL** Use tree diagrams to determine all possible outcomes.

Problem

What is the probability of a coin landing Heads exactly two times in three tosses?

Follow these instructions to use a tree diagram to calculate the probability.

A. Suppose you tossed a coin twice.
Write the possible outcomes for the first toss.

_____ and _____

Write the possible outcomes for the second toss.

_____ and _____

B. Complete the tree diagram for the two tosses.
H represents Heads and T represents Tails.

1st toss	2nd toss	Outcome
H	H	H–H
	T	H–T
T	___	___–___
	___	___–___

C. Write the possible outcomes for the third toss.

_____ and _____

D. Extend the tree diagram for the third toss.

1st toss	2nd toss	3rd toss	Outcome

H → H → H → H–H–H

H → H → T → H–H–T

H → T → H → __ – __ – __

H → T → __ – __ – __

T → __ → __ – __ – __

T → __ → __ – __ – __

T → __ → __ – __ – __

T → __ → __ – __ – __

Hint

Follow the arrows to find each outcome.

E. Circle the outcomes that show exactly two Heads in three tosses.

F. What is the probability of tossing exactly two Heads in three tosses?

$$P(\text{2 Heads}) = \frac{\square}{\square}$$

G. If the coin were tossed a fourth time, how many outcomes would there be? _____

How do you know?

_____

_____

Reflecting

▶ How does a tree diagram help you list all the possible outcomes?

_____

_____

_____

_____

Practising

Text page 428

4. Joanna sells irises and lilies in bunches of four. She randomly selects the flowers.

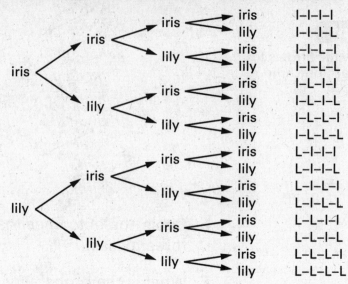

1st flower	2nd flower	3rd flower	4th flower	Outcome

a) Circle the outcomes that show one iris and three lilies.

b) What is the probability that a bunch will have one iris and three lilies?

$$P(1 \text{ iris and } 3 \text{ lilies}) = \frac{\square}{\square}$$

6. **a)** How many possible outcomes are there each time you roll a single die? _____

b) Complete the tree diagram for two rolls of a die. How many possible outcomes are there? _____

1st roll	2nd roll	Sum

c) Complete the Sum column.

d) (Circle) the outcomes that total a sum of 7.

e) What is the probability of tossing a sum of 7 using two rolls of a die?

$$P(\text{sum of 7}) = \frac{\square}{\square}$$

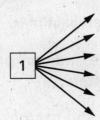

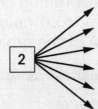

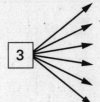

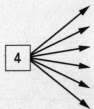

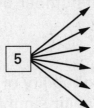

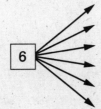

12.5 Applying Probabilities

Text page 430

You will need
- a calculator
- two dice

▶ **GOAL** Calculate and compare probabilities.

Hint

In the table, consecutive means the second die is either one greater or one less than the first die; for example, 5-6 or 4-3.

Use these steps to calculate and compare theoretical and experimental probabilities.

Step 1: The tree diagram in the margin displays the possible outcomes for rolling two dice.
Use the tree diagram to complete the Number of outcomes column of the table.

1st die	2nd die	Outcome
1	1	1–1
	2	1–2
	3	1–3
	4	1–4
	5	1–5
	6	1–6
2	1	2–1
	2	2–2
	3	2–3
	4	2–4
	5	2–5
	6	2–6
3	1	3–1
	2	3–2
	3	3–3
	4	3–4
	5	3–5
	6	3–6
4	1	4–1
	2	4–2
	3	4–3
	4	4–4
	5	4–5
	6	4–6
5	1	5–1
	2	5–2
	3	5–3
	4	5–4
	5	5–5
	6	5–6
6	1	6–1
	2	6–2
	3	6–3
	4	6–4
	5	6–5
	6	6–6

Theoretical Probability			
Result	**Number of outcomes**	**Probability**	
		Fraction	**Percent**
both even or both odd			
consecutive			
same number			
none of the above results			

Step 2: What is the total number of possible outcomes?

_____

Calculate the probability of each result and write each probability in the table above as a fraction and as a percent.

Step 3: Roll two dice 20 times. Tally each result in the table below as you roll.

Experimental Probability			
Result	Tally of outcomes	Probability	
		Fraction	Percent
both even or both odd			
consecutive			
same number			
none of the above results			

Step 4: How many outcomes are there in your table?

_____

Calculate the probability of each result and write each probability in the table above as a fraction and as a percent.

Step 5: How does your experimental probability in the table above compare to the theoretical probability in the table on page 278?

_____

_____

Reflecting

▶ The probability of each result can be written as a fraction, a decimal, or a percent. Which form makes comparing the probabilities easiest? Give reasons for your answer.

_____

_____

Practising

Text page 432

3. Consider the following statistics:

- Raj's batting average is .570.
- Bella gets a hit 3 out of every 5 times at bat.
- Connie has a 58% chance of getting a hit.
- For Derek, $P(\text{hit}) = \frac{2}{3}$.

a) Write each probability as a percent in the table below.

Player	Batting stat	Probability (percent)
Raj	.570	
Bella	3 out of 5	
Connie	58%	
Derek	$\frac{2}{3}$	

b) Which player has the greatest probability of getting a hit? _____

8. To play the birthday wheel, you pay $1. Then you spin the wheel. If your birthday month comes up, you win $10.

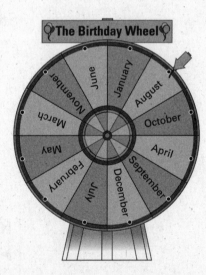

a) How many possible outcomes are there on the wheel? _____

b) What is the probability that you will win on one spin?

$$P(\text{win}) = \frac{\Box}{\Box}$$

10. You have three bags of marbles.

- Bag A has 3 black marbles and 5 white marbles.
- Bag B has 8 black marbles and 10 white marbles.
- Bag C has 17 black marbles and 20 white marbles.

a) Complete the Fraction columns of the table by calculating the probability of drawing black and white marbles from each bag.

Bag	Probability of drawing black		Probability of drawing white	
	Fraction	Decimal	Fraction	Decimal
A	$\frac{}{8}$			
B			$\frac{}{18}$	
C				

b) Use your calculator to express each probability as a decimal. Complete the Decimal columns of the table.

c) Which bag has the greatest probability of drawing a black marble? _____

d) Which bag has the greatest probability of drawing a white marble? _____

e) If all the marbles are placed in one bag, there would be

_____ black marbles
_____ white marbles
_____ total marbles

What is the probability of drawing a white marble from this bag? Write the probability as a fraction and as a decimal.

$P(\text{white marble}) = \dfrac{\boxed{}}{\boxed{}} = $ _____

Parallelogram A

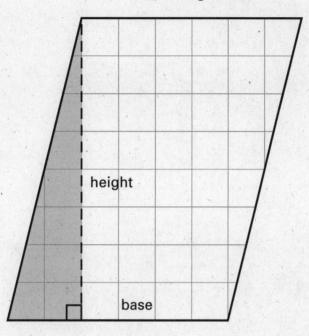

height

base

Parallelogram B

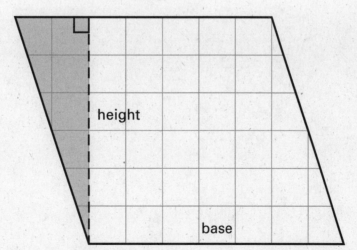

height

base

△**ABC**

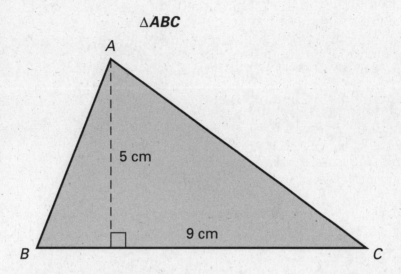

△**ABC**

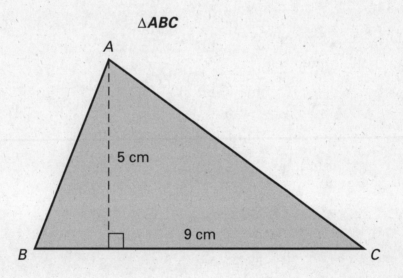

△**DEF**   △**DEF**

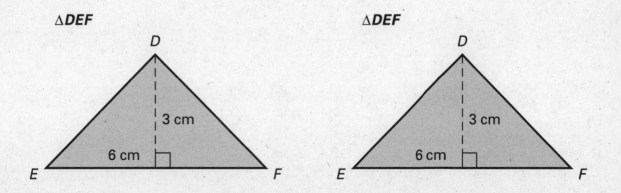

Trapezoid *ABCD*

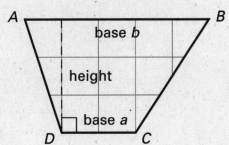

Trapezoid *ABCD*

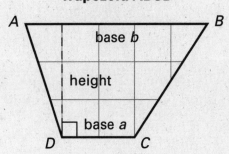

Trapezoid *DEFG*

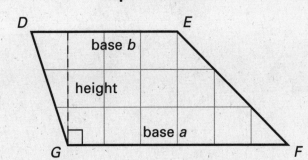

Trapezoid *DEFG*

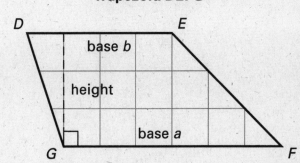

T pentomino

L pentomino

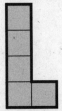

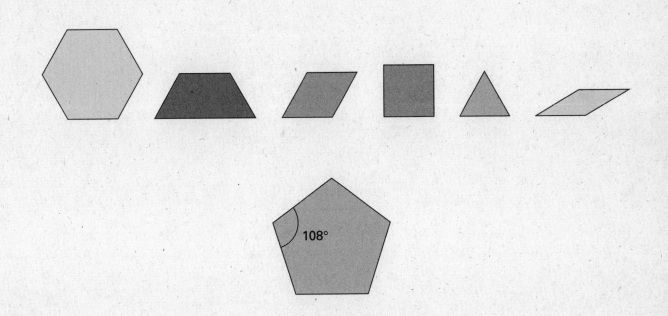

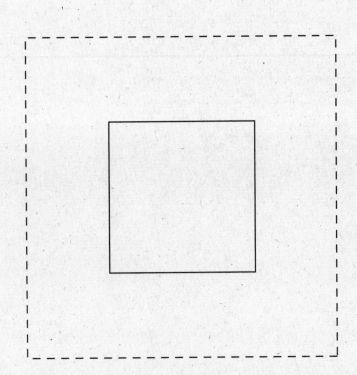

Strip #1			

Strip #2		

Strip #3											

Cutout Page 10.2

1

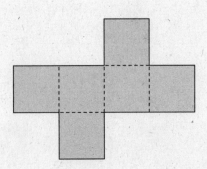

2

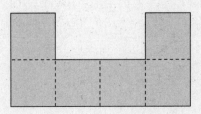

3

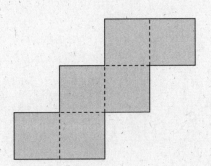

4

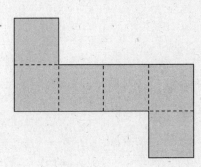

Question 8 a)

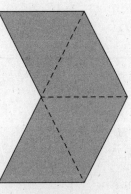
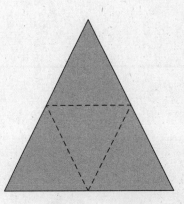

Question 8 b)

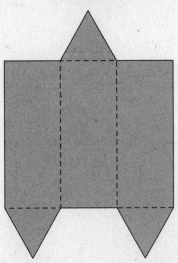
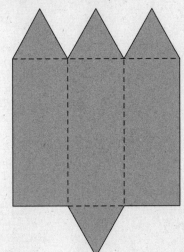

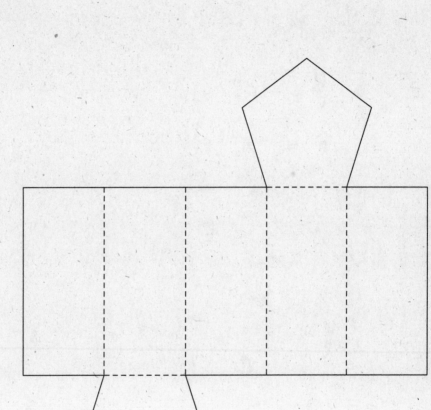